Mathematical Modeling using Differential Equations, and Network Theory

Mathematical Modeling using Differential Equations, and Network Theory

Special Issue Editor

Ioannis Dassios

MDPI • Basel • Beijing • Wuhan • Barcelona • Belgrade • Manchester • Tokyo • Cluj • Tianjin

Special Issue Editor
Ioannis Dassios
University College Dublin
Ireland

Editorial Office
MDPI
St. Alban-Anlage 66
4052 Basel, Switzerland

This is a reprint of articles from the Special Issue published online in the open access journal *Applied Sciences* (ISSN 2076-3417) (available at: https://www.mdpi.com/journal/applsci/special_issues/network_theory).

For citation purposes, cite each article independently as indicated on the article page online and as indicated below:

LastName, A.A.; LastName, B.B.; LastName, C.C. Article Title. *Journal Name* **Year**, *Article Number*, Page Range.

ISBN 978-3-03928-825-0 (Hbk)
ISBN 978-3-03928-826-7 (PDF)

Contents

About the Special Issue Editor

Ioannis Dassios is currently a UCD Research Fellow/Assistant Professor at AMPSAS, University College Dublin, Ireland. His research interests include dynamical systems, mathematics of networks, differential and difference equations, singular systems, fractional calculus, optimization methods, linear algebra, and mathematical modeling (materials, electrical power systems, economic models, etc). He studied Mathematics, completed a two-year MSc in Applied Mathematics and Numerical Analysis, and obtained his Ph.D. degree in Applied Mathematics at University of Athens, Greece with the grade "Excellent" (highest mark in the Greek system). He had positions at the University of Edinburgh, U.K; University of Manchester, U.K.; and University of Limerick, Ireland. He has published more than 55 articles in internationally leading academic journals and has participated in several international collaborations. He has served as a reviewer more than 500 times in more than 75 different journals, he has been member in scientific and organizing committees of international conferences, and he is also member of editorial boards of peer-reviewed journals. Finally, he has received several awards such as travel grants, for reviewing and for his contributions to his institute.

applied sciences

MDPI

Editorial

Special Issue on Mathematical Modeling Using Differential Equations and Network Theory

Ioannis Dassios

AMPSAS, University College Dublin, Dublin 4, Ireland; ioannis.dassios@ucd.ie

Received: 1 March 2020; Accepted: 3 March 2020; Published: 10 March 2020

1. Introduction

This special issue collects the latest results on differential/difference equations, the mathematics of networks, and their applications to engineering, and physical phenomena. The Special Issue has 42 submissions and eight high-quality papers which got published with original research results. The Special Issue brought together mathematicians with physicists, engineers, as well as other scientists. Topics covered in this issue:

- Differential/difference equations
- Mathematics of networks
- Fractional calculus
- Partial differential equations
- Discrete calculus
- Mathematical models using dynamical systems

2. Acoustic Wave Equations Using Fractional-Order Differential Equations

In [1], the authors present a newly developed technique, defined as a variational homotopy perturbation transform method in order to solve fractional-order acoustic wave equations. The basic idea behind this article is to extend the variational homotopy perturbation method to the variational homotopy perturbation transform method.

The proposed method is an accurate and straightforward technique to solve fractional-order partial differential equations, and can be considered as a practical analytical technique to solve non-linear fractional partial differential equations compared to other analytical techniques existing in the literature. Several illustrative examples verify the method.

3. Analytical Solutions of Dimensional Physical Models Using Modified Decomposition Method

In [2], the authors present a new analytical technique based on an innovative transformation in order to solve (2+time fractional-order) dimensional physical models. The proposed method is based on the hybrid methodology of Shehu transformation along with the Adomian decomposition method.

The solutions of the targeted problems are represented by graphs and are obtained in a series form that has the desired rate of convergence. The method is, in general, a practical analytical technique to solve linear and non-linear fractional partial differential equations. Numerical examples are given using the proposed method.

4. Multi-Switching Combination Synchronization of Fractional-Order Delayed Systems

In [3] the authors investigate multi-switching combination synchronization of three fractional-order delayed systems. This is actually a generalization of previous multi-switching combination synchronization of fractional-order systems by introducing time-delays.

Based on the stability theory of linear fractional-order systems with multiple time-delays, the article provides appropriate controllers to obtain multi-switching combination synchronization of three non-identical fractional-order delayed systems. In addition, numerical simulations show that they are in accordance with the theoretical analysis given.

5. An Overview of Early Developments of the Hardy–Cross-Type Methods

In [4], the authors provide an overview of early developments of the Hardy–Cross-type methods for computation of flow distribution in pipe networks.

Cross originally proposed a method for analysis of flow in networks of conduits or conductors in 1936. His method was the first really useful engineering method in the field of pipe network calculation. Only electrical analogs of hydraulic networks were used before the Hardy–Cross method. A problem with flow resistance versus electrical resistance makes these electrical analog methods obsolete. The method by Hardy–Cross is taught extensively at faculties, and it remains an important tool for the analysis of looped pipe systems. Engineers today mostly use a modified Hardy–Cross method that considers the whole looped network of pipes simultaneously (use of these methods without computers is practically impossible).

In addition, in [4] a method from a Russian practice published during the 1930s, which is similar to the Hardy–Cross method, is also described. Some notes from the work of Hardy–Cross are also presented. Furthermore, an improved version of the Hardy–Cross method, which significantly reduces the number of iterations, is presented and discussed.

Finally, the authors present results on tested multi-point iterative methods, which can be used as a substitution for the Newton–Raphson approach used by Hardy–Cross.

6. Parametrical Non-Complex Tests to Evaluate Partial Decentralized Linear-Output Feedback Control Stabilization Conditions

In [5], the authors formulate sufficiency-type linear-output feedback decentralized closed-loop stabilization conditions if the continuous-time linear dynamic system can be stabilized under linear output-feedback centralized stabilization.

The provided tests are simple to evaluate, while they are based on the quantification of the sufficient smallness of the parametrical error norms between the control, output, interconnection and open-loop system dynamics matrices and the corresponding control gains in the decentralized case related to the centralized counterpart.

The tolerance amounts of the various parametrical matrix errors are described by the maximum allowed tolerance upper-bound of a small positive real parameter that upper-bounds the various parametrical error norms. Such a tolerance is quantified by considering the first or second powers of such a small parameter.

The results are seen to be directly extendable to quantify the allowed parametrical errors that guarantee the closed-loop linear output-feedback stabilization of a current system related to its nominal counterpart. Several numerical examples are included and discussed in the article.

7. Transient-Flow Modeling of Vertical Fractured Wells with Multiple Hydraulic Fractures

Massive hydraulic fracturing of vertical wells has been extensively employed in the development of low-permeability gas reservoirs. The existence of multiple hydraulic fractures along a vertical well makes the pressure profile around the vertical well complex.

In [6], the authors study the pressure dependence of permeability in order to develop a seepage model of vertically fractured wells with multiple hydraulic fractures. Both transformed pseudo-pressure and perturbation techniques have been employed to linearize the proposed model.

The proposed work further enriches the understanding of the influence of the stress sensitivity on the performance of a vertical fractured well with multiple hydraulic fractures and can be used to more accurately interpret and forecast the transient pressure.

Some key points in the article are the superposition principle and a hybrid analytical-numerical method that are used to obtain the bottom-hole pseudo-pressure solution, the type curves for pseudo-pressure that are presented and identified, and finally, the discussion that is included on the effects of the relevant parameters on the type curve and the error caused by neglecting the stress sensitivity.

8. Policy-Compliant Maximum Network Flows

Computer network administrators are often interested in the maximal bandwidth that can be achieved between two nodes in the network, or how many edges can fail before the network gets disconnected. Classic maximum flow algorithms that solve these problems are well-known. However, in practice, network policies are in effect, severely restricting the flow that can actually be set up. These policies are put into place to conform to service level agreements and optimize network throughput, and can have a large impact on the actual routing of the flows.

In [7], the authors model the problem and define a series of progressively more complex conditions and algorithms that calculate increasingly tighter bounds on the policy-compliant maximum flow using regular expressions and finite-state automata. This is the first time that specific conditions are deduced, which characterize how to calculate policy-compliant maximum flows using classic algorithms on an unmodified network.

9. The Fractional Form of the Tinkerbell Map Is Chaotic

In [8], the authors are concerned with a fractional Caputo-difference form of the well-known Tinkerbell chaotic map. The dynamics of the proposed map are investigated numerically through phase plots, bifurcation diagrams, and Lyapunov exponents considered from different perspectives.

In addition, a stabilization controller is proposed, and the asymptotic convergence of the states is established by means of the stability theory of linear fractional discrete systems. Numerical results are employed to confirm the analytical findings.

Funding: This Editorial was supported by: Science Foundation Ireland, by funding Ioannis Dassios under Grant No. SFI/15/IA/3074.

Acknowledgments: This issue would not have been possible without the help of a variety of talented authors, professional reviewers, and the dedicated editorial team of Applied Sciences. Thank you to all the authors and reviewers for this opportunity. Finally, thanks to the Applied Sciences editorial team.

Conflicts of Interest: The author declares no conflict of interest.

References

1. Ali, I.; Khan, H.; Shah, R.; Baleanu, D.; Kumam, P.; Arif, M. Fractional View Analysis of Acoustic Wave Equations, Using Fractional-Order Differential Equations. *Appl. Sci.* **2020**, *10*, 610. [CrossRef]
2. Khan, H.; Farooq, U.; Shah, R.; Baleanu, D.; Kumam, P.; Arif, M. Analytical Solutions of (2+Time Fractional Order) Dimensional Physical Models, Using Modified Decomposition Method. *Appl. Sci.* **2020**, *10*, 122. [CrossRef]
3. Li, B.; Wang, Y.; Zhou, X. Multi-Switching Combination Synchronization of Three Fractional-Order Delayed Systems. *Appl. Sci.* **2019**, *9*, 4348. [CrossRef]
4. Brkić, D.; Praks, P. Short Overview of Early Developments of the Hardy Cross Type Methods for Computation of Flow Distribution in Pipe Networks. *Appl. Sci.* **2019**, *9*, 2019. [CrossRef]
5. De la Sen, M.; Ibeas, A. Parametrical Non-Complex Tests to Evaluate Partial Decentralized Linear-Output Feedback Control Stabilization Conditions from Their Centralized Stabilization Counterparts. *Appl. Sci.* **2019**, *9*, 1739. [CrossRef]
6. Guo, P.; Sun, Z.; Peng, C.; Chen, H.; Ren, J. Transient-Flow Modeling of Vertical Fractured Wells with Multiple Hydraulic Fractures in Stress-Sensitive Gas Reservoirs. *Appl. Sci.* **2019**, *9*, 1359. [CrossRef]

7. Audenaert, P.; Colle, D.; Pickavet, M. Policy-Compliant Maximum Network Flows. *Appl. Sci.* **2019**, *9*, 863. [CrossRef]
8. Ouannas, A.; Khennaoui, A.; Bendoukha, S.; Vo, T.; Pham, V.; Huynh, V. The Fractional Form of the Tinkerbell Map Is Chaotic. *Appl. Sci.* **2018**, *8*, 2640. [CrossRef]

applied sciences

MDPI

Article

The Fractional Form of the Tinkerbell Map Is Chaotic

Adel Ouannas [1], Amina-Aicha Khennaoui [2], Samir Bendoukha [3], Thoai Phu Vo [4], Viet-Thanh Pham [5,*] and Van Van Huynh [6]

1 Department of Mathematics and Computer Science, University of Larbi Tebessi, Tebessa 12002, Algeria; adel.ouannas@yahoo.com
2 Department of Mathematics and Computer Sciences, University of Larbi Ben M'hidi, Oum El Bouaghi 04000, Algeria; kamina_aicha@yahoo.fr
3 Electrical Engineering Department, College of Engineering at Yanbu, Taibah University, Medina 42353, Saudi Arabia; sbendoukha@taibahu.edu.sa
4 Faculty of Electrical and Electronics Engineering, Ton Duc Thang University, Ho Chi Minh City, Vietnam; vophuthoai@tdtu.edu.vn
5 Nonlinear Systems and Applications, Faculty of Electrical and Electronics Engineering, Ton Duc Thang University, Ho Chi Minh City, Vietnam
6 Modeling Evolutionary Algorithms Simulation and Artificial Intelligence, Faculty of Electrical and Electronics Engineering, Ton Duc Thang University, Ho Chi Minh City, Vietnam; huynhvanvan@tdtu.edu.vn
* Correspondence: phamvietthanh@tdtu.edu.vn

Received: 3 December 2018; Accepted: 12 December 2018; Published: 16 December 2018

Abstract: This paper is concerned with a fractional Caputo-difference form of the well-known Tinkerbell chaotic map. The dynamics of the proposed map are investigated numerically through phase plots, bifurcation diagrams, and Lyapunov exponents considered from different perspectives. In addition, a stabilization controller is proposed, and the asymptotic convergence of the states is established by means of the stability theory of linear fractional discrete systems. Numerical results are employed to confirm the analytical findings.

Keywords: fractional discrete calculus; discrete chaos; Tinkerbell map; bifurcation; stabilization

1. Introduction

Throughout the last 50 years, chaotic dynamical systems have attracted increasing attention due to their applicability in a range of diverse and multidisciplinary fields. A dynamical system is said to be chaotic if its states are extremely sensitive to small variations in the initial conditions. Another important property of chaotic systems is that they have attractors characterized by a complicated set of points with a fractal structure commonly referred to as a strange attractor. This chaotic behavior was first observed in continuous dynamical systems and was thought to be an undesirable property. The first chaotic system encountered in the modeling of a real-life phenomena is that of Lorenz [1], which describes atmospheric convection. Soon after, researchers found that chaotic systems can also be discrete. A number of chaotic maps were proposed throughout the years including the Hénon map [2], the logistic map [3], the Lozi map [4], the 3D Stefanski map [5], the Rössler map [6], and many more. Recently, nonlinear oscillations on Riemannian manifolds that can exhibit a chaotic behavior were introduced in [7,8]. Other related works include an investigation of the chaotic dynamics in a fractional love model with an external environment, as in [9], and an extension using a fuzzy function [10].

In recent years, with the growing interest in fractional discrete calculus [11], people have started looking into fractional chaotic maps. Although fractional maps come with considerable added complexity, they provide better flexibility in the modeling of natural phenomena and lead to richer dynamics with more degrees of freedom. Among the fractional chaotic maps that have been proposed, studied, and applied over the last five years are the fractional logistic map [12], the fractional Hénon map [13], the generalized hyperchaotic Hénon map [14], and the fractional unified map [15]. Perhaps the main concern of the research community has been the possibility of controlling and synchronizing these types of maps [15–20]. An application of a generalized fractional logistic map to data encryption and its FPGA implementation was achieved in [21].

In this paper, we are interested in the Tinkerbell discrete-time chaotic system, which is of the form:

$$\begin{cases} x(n+1) = x^2(n) - y^2(n) + \alpha x(n) + \beta y(n), \\ y(n+1) = 2x(n)y(n) + \gamma x(n) + \delta y(n), \end{cases} \tag{1}$$

where α, β, γ, and δ are system parameters and n represents the discrete iteration step. It is rumored that the map (1) derives its name from the famous Cinderella story, as the trajectory followed by the map resembles that of Tinkerbell appearing in the movie adaptation of the fairy tale. The Tinkerbell map has been studied by many as it exhibits very rich dynamics including a chaotic behavior and a range of periodic states. For instance, its bifurcation subject to different scenarios and initial settings has been studied in [22–25]. A more comprehensive study was performed in [26]. The authors identified conditions for the existence of fold bifurcation, flip bifurcation, and Hopf bifurcation in the Tinkerbell map.

In order to visualize the dynamics of the map (1), we resort to phase plots, bifurcation diagrams, and Lyapunov exponent estimation. We assume parameter values $\alpha = 0.9$, $\beta = -0.6013$, $\gamma = 2$, and $\delta = 0.5$ and initial states $(x(0), y(0)) = (-0.72, -0.64)$. The results are depicted in Figure 1. The Tinkerbell map's phase plot is depicted in Figure 1a. Based on Figure 1b, we can see that the estimated Lyapunov exponents of (1) are given by $\lambda_1 \approx 0.2085$ and $\lambda_2 \approx -0.4925$. It is well known that a positive Lyapunov exponent indicates a chaotic behavior. The remaining parts of Figure 1 depict the bifurcation diagrams of the map (1) with respect to different parameters. These diagrams confirm that the map exhibits a range of different behaviors.

It should be clear to the reader that the Tinkerbell map has rich dynamics and is heavily dependent on its parameters, as well as the initial setting. The main objective of this paper is to investigate the fractional Caputo-difference form of the Tinkerbell map in order to benefit from the added degrees of freedom due to the fractional nature. It is expected that the fractional form will have even richer dynamics and may consequently be more suitable for applications that require a higher entropy level such as data/image encryption.

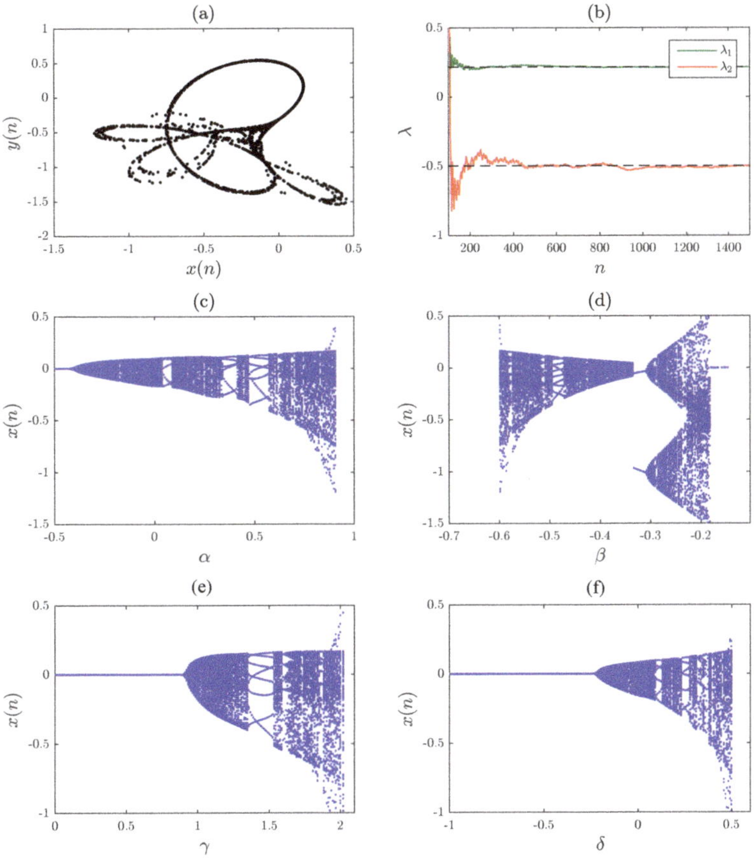

Figure 1. (**a**) Attractor of the Tinkerbell map (2) with $(\alpha, \beta, \gamma, \delta) = (0.9, -0.6013, 2, 0.5)$ and initial conditions $(x(0), y(0)) = (-0.72, -0.64)$. (**b**) Estimated Lyapunov exponents by means of the Jacobian matrix method. (**c**) Bifurcation plot with $\alpha \in [-0.5, 1]$ as the critical parameter and $\Delta\alpha = 0.0075$. (**d**) Bifurcation plot with $\beta \in [-0.6, -0.1]$ as the critical parameter and $\Delta\beta = 0.0025$. (**e**) Bifurcation plot with $\gamma \in [0, 2.1]$ as the critical parameter and $\Delta\gamma = 0.01$. (**f**) Bifurcation plot with $\delta \in [-1, 0.6]$ as the critical parameter and $\Delta\delta = 0.008$.

2. Fractional Tinkerbell Map

In this section, we use recent developments in fractional discrete calculus to define the Caputo-difference fractional map corresponding to (1). First, let us define the vth fractional sum of anarbitrary function $X(t)$ [27] as:

$$\Delta_a^{-v} X(t) = \frac{1}{\Gamma(v)} \sum_{s=a}^{t-v} (t-s-1)^{(v-1)} X(s), \tag{2}$$

for $t \in \mathbb{N}_{a+n-v}$ and $v > 0$, where $\mathbb{N}_a := \{a, a+1, a+2, ...\}$. Note that the term $t^{(v)}$ is known as the falling function and may be defined by means of the Gamma function Γ as:

$$t^{(v)} = \frac{\Gamma(t+1)}{\Gamma(t+1-v)}. \tag{3}$$

Based on this definition of the fractional sum, we may define the Caputo-like fractional difference operator.

In this section, we would like to produce a fractional difference form of the Tinkerbell map (1). First, we take the difference form, which for function $x(t) : \mathbb{N}_a \to \mathbb{R}$ with fractional order $v \notin \mathbb{N}$ is given by:

$$^{C}\Delta_a^v x(t) = \Delta_a^{-(n-v)} \Delta^n x(t). \tag{4}$$

Substituting yields the final form proposed in [28], which is defined as:

$$^{C}\Delta_a^v x(t) = \frac{1}{\Gamma(n-v)} \sum_{s=a}^{t-(n-v)} (t-s-1)^{(n-v-1)} \Delta^n x(s), \tag{5}$$

where $t \in \mathbb{N}_{a+n-v}$ and $n = \lceil v \rceil + 1$.

We are now ready to examine the fractional map. First, we take the difference form of (1) to obtain:

$$\begin{cases} \Delta x(n) = x^2(n) - y^2(n) + (\alpha - 1)x(n) + \beta y(n), \\ \Delta y(n) = 2x(n)y(n) + \gamma x(n) + (\delta - 1)y(n). \end{cases} \tag{6}$$

We may replace the standard difference in (6) with the Caputo-difference, which yields:

$$\begin{cases} ^{C}\Delta_a^v x(t) = x^2(t-1+v) - y^2(t-1+v) \\ \qquad + (\alpha-1)x(t-1+v) + \beta y(t-1+v), \\ ^{C}\Delta_a^v y(t) = 2x(t-1+v)y(t-1+v) + \gamma x(t-1+v) \\ \qquad + (\delta-1)y(t-1+v), \end{cases} \tag{7}$$

for $t \in \mathbb{N}_{a+1-v}, 0 < v \leq 1$, a is the starting point, and $^{C}\Delta_a^v$ is a Caputo-like difference operator. The case $v = 1$ corresponds to the non-fractional scenario (1).

3. Dynamics of the Fractional Tinkerbell Map

In this section, we will employ numerical tools to assess the dynamics of the proposed fractional Tinkerbell map (7). For that, we will need a discrete numerical formula that allows us to evaluate the states of the map in fractional discrete time. According to [29] and other similar studies, we can evaluate (7) numerically as:

$$\begin{cases} x(n) = x(0) + \frac{1}{\Gamma(v)} \sum_{j=1}^{n} \frac{\Gamma(n-j+v)}{\Gamma(n-j+1)} \left[x^2(j-1) - y^2(j-1) \right. \\ \qquad \left. + (\alpha-1)x(j-1) + \beta y(j-1)\right], \\ y(n) = y(0) + \frac{1}{\Gamma(v)} \sum_{j=1}^{n} \frac{\Gamma(n-j+v)}{\Gamma(n-j+1)} \left[2x(j-1)y(j-1) \right. \\ \qquad \left. + \gamma x(j-1) + (\delta-1)y(j-1)\right], \end{cases} \tag{8}$$

where we assumed $a = 0$ for simplicity. This yields an initial-value problem similar to that of [30], which allows us to use a similar discrete integral equation.

Using Formula (8), we may obtain the states of the fractional Tinkerbell map and consequently produce time series plots of the states, phase-space plots, and bifurcation diagrams. We start with a simple case where the parameters and initial conditions are identical to those adopted in the standard case, i.e., $(\alpha, \beta, \gamma, \delta) = (0.9, -0.6013, 2, 0.5)$ and $(x(0), y(0)) = (-0.72, -0.64)$. Given the fractional order $v = 0.98$, Figure 2 depicts the discrete time evolution of the states. Since the time series in Figure 2 do not indicate the existence or absence of chaos definitively, it is more convenient to show the trajectories followed by the map in state space. Figure 3 shows the phase plots for different values of the fractional order $v \in \{0.995, 0.99, 0.97, 0.952\}$. We see that the overall Tinkerbell shape remains valid for a short range of fractional orders. As the order gets close to 0.95, the trajectory almost completely disappears.

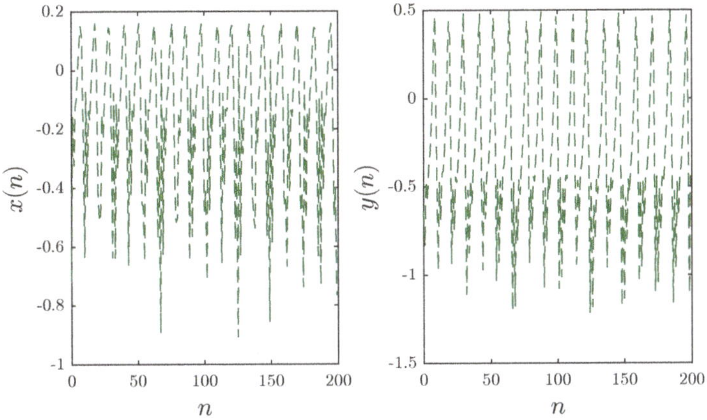

Figure 2. Time evolution of the fractional Tinkerbell map's states with parameters $(\alpha, \beta, \gamma, \delta) = (0.9, -0.6013, 2, 0.5)$, initial conditions $(x(0), y(0)) = (-0.72, -0.64)$, and fractional order $v = 0.98$.

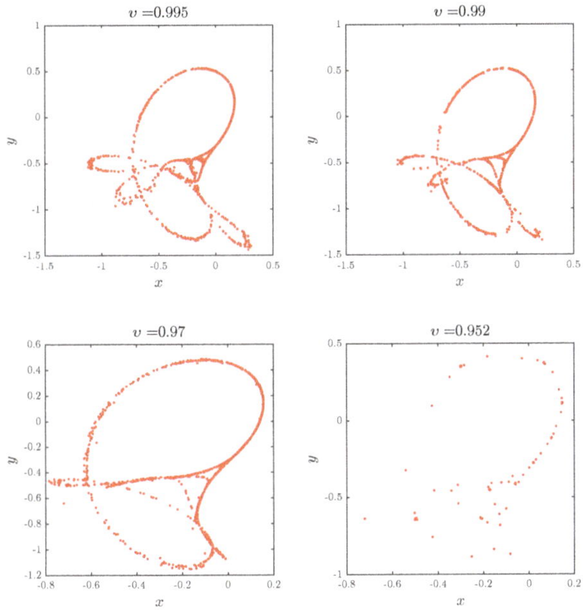

Figure 3. Phase plots of the fractional Tinkerbell map (7) for parameters $(\alpha, \beta, \gamma, \delta) = (0.9, -0.6013, 2, 0.5)$, initial conditions $(x(0), y(0)) = (-0.72, -0.64)$, and different fractional orders.

Although the phase plots give an indication of the behavior of the map, it is not until we visualize the bifurcation of the map subject to different parameters that a more complete picture forms. We choose the parameter β as the critical parameter and varied it over the range $\beta \in [-0.6, -0.1]$ in steps of $\Delta\beta = 0.0025$. The process may be easily repeated for other parameters. The bifurcation diagrams obtained using the same parameter and initial condition values from earlier are depicted in Figure 4. We observe that although the general dynamics remain similar, the intervals seem to become shorter as the fractional order is decreased.

Even though these bifurcation diagrams suggest the existence of chaos in the fractional Tinkerbell map, they are not definitive. Generally, in order to prove the existence of chaos, we must use multiple tools including time series, phase portraits, Poincaré maps, power spectra, bifurcation diagrams, Lyapunov exponents, etc. The next tool at our disposal is Lyapunov exponents. We calculate these exponents by means of the Jacobian method. It is well known that when λ_{max} is positive and the points in the corresponding bifurcation diagram are dense, the map is highly likely to be chaotic. Figure 5 shows the largest Lyapunov exponents corresponding to the same bifurcation diagrams depicted in Figure 4 in the x-β plane. We can observe clearly that for certain ranges of the parameter β, chaos exists.

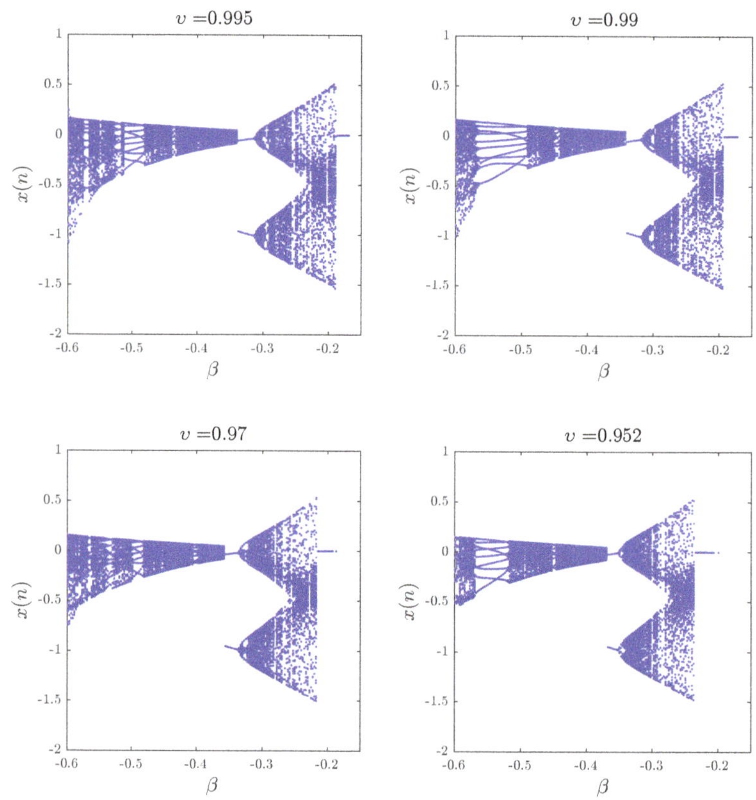

Figure 4. Bifurcation diagrams of the fractional Tinkerbell map (7) with $\beta \in [-0.6, -0.1]$ being changed in steps of $\Delta\beta = 0.0025$, parameters $(\alpha, \gamma, \delta) = (0.9, 2, 0.5)$, initial conditions $(x(0), y(0)) = (-0.72, -0.64)$, and different fractional orders.

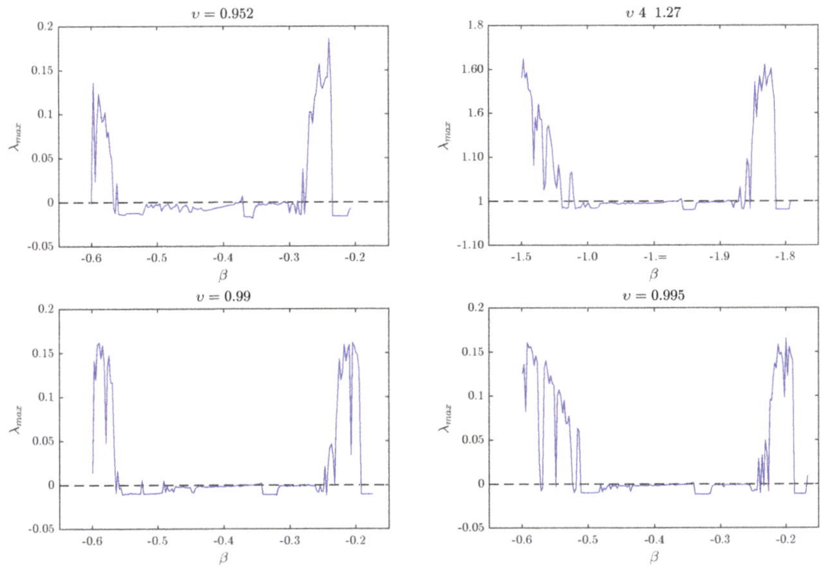

Figure 5. The largest Lyapunov exponent as a function of parameter β for different values of the fractional order.

Another interesting aspect is the effect of the fractional order on the dynamics of the map for a specific set of parameter values. We fix the parameters and initial conditions at $(\alpha, \beta, \gamma, \delta) = (0.9, -0.6013, 2, 0.5)$ and $(x(0), y(0)) = (-0.72, -0.64)$, respectively. Figure 6 shows the bifurcation plot with the critical parameter $v \in [0, 1]$ being changed in steps of $\Delta v = 0.005$. This is interesting in that it shows that although the chaotic behavior disappears when the fractional order drops close to 0.95, it is observed again over intermittent intervals. Chaos does not disappear totally until the fractional order is very low.

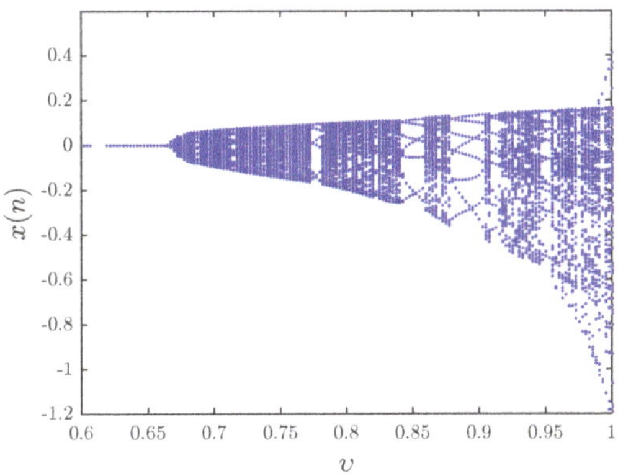

Figure 6. Bifurcation diagram of the fractional Tinkerbell map (7) with $v \in [0, 1]$, $\Delta v = 0.005$, $(\alpha, \beta, \gamma, \delta) = (0.9, -0.6013, 2, 0.5)$, and $(x(0), y(0)) = (-0.72, -0.64)$.

The largest Lyapunov exponent corresponding to the this bifurcation diagram in the x-v plane is depicted in Figure 7. From the figure, we observe that for a fractional order larger than 0.952, λ_{max} is positive, which implies that the fractional Tinkerbell map is chaotic. During the interval $(0.6609, 0.952)$,

λ_{max} is observed to change intermittently between positive and negative signs, which means that chaos starts to appear and disappear. Finally, for values lower than 0.6609, chaos disappears completely. These results agree with the bifurcation diagram in Figure 6.

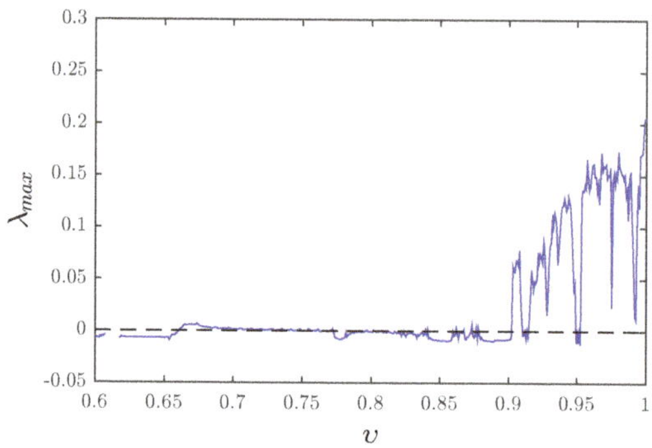

Figure 7. The largest Lyapunov exponent as a function of the fractional order v for the same parameters and initial conditions in Figure 6.

4. Control of the Fractional Tinkerbell Map

In this section, we show that the proposed fractional Tinkerbell can be stabilized by means of a simple adaptive feedback controller. In order to be able to establish the asymptotic convergence of the controlled states towards zero, we first need to recall some important results from the literature concerning the asymptotic stability of fractional discrete systems. Since fractional discrete calculus is still relatively new, the existing literature related to stability is very limited. There are two main ways of establishing asymptotic stability. The first relies on the linearity of the system and places conditions on the eigenvalues of the Jacobian [31]. The second scheme is a generalization of the well-known Lyapunov direct method [32]. Although, the Lyapunov method is powerful and can support different types of systems, its has yet to be established for delayed fractional discrete systems, which renders it unusable for the system at hand. Hence, our objective here is to design the control laws to linearize the system, which will allow us to use the stability theory of linear systems. The following theorem summarizes the result of [31].

Theorem 1. *The zero equilibrium of the linear fractional discrete system:*

$$ {}^{C}\Delta_{a}^{v} F(t) = MF(t+v-1), \tag{9}$$

where $F(t) = (f_1(t), ..., f_n(t))^T$, $0 < v \leq 1$, *and* $M \in \mathbb{R}^{n\times n}$, *is asymptotically stable if the eigenvalues* λ *of* M *satisfy:*

$$\lambda \in \left\{ z \in \mathbb{C} : \ |z| < \left(2\cos\frac{|\arg z| - \pi}{2-v} \right)^{v} \ \text{and} \ |\arg z| > \frac{v\pi}{2} \right\} \tag{10}$$

for all $t \in \mathbb{N}_{a+1-v}$.

Consider the controlled version of (7) given by:

$$\begin{cases} {}^{C}\Delta_{a}^{v} x(t) = x^2(t-1+v) - y^2(t-1+v) + (\alpha-1)x(t-1+v) \\ \qquad\qquad +\beta y(t-1+v) + u_x(t-1+v), \\ {}^{C}\Delta_{a}^{v} y(t) = 2x(t-1+v)y(t-1+v) + \gamma x(t-1+v) \\ \qquad\qquad +(\delta-1)y(t-1+v) + u_y(t-1+v), \end{cases} \tag{11}$$

where $u_x(t)$ and $u_y(t)$ are adaptive control terms. The following theorem presents the proposed control laws.

Theorem 2. *The states of the controlled 2D fractional Tinkerbell map (11) are guaranteed to converge towards zero asymptotically subject to:*

$$\begin{cases} u_x(t) = y^2(t) - x^2(t), \\ u_y(t) = -2x(t)\,y(t) - \gamma x(t). \end{cases} \tag{12}$$

Proof. Substituting (12) into (11) yields the dynamics:

$$\begin{cases} {}^{C}\Delta_a^v x(t) = (\alpha - 1)\,x(t-1+v) + \beta y(t-1+v), \\ {}^{C}\Delta_a^v y(t) = (\delta - 1)\,y(t-1+v), \end{cases} \tag{13}$$

or more compactly:

$${}^{C}\Delta_a^v (x(t), y(t))^T = A\,(x(t), y(t))^T, \tag{14}$$

with:

$$A = \begin{pmatrix} \alpha - 1 & \beta \\ 0 & \delta - 1 \end{pmatrix}. \tag{15}$$

The eigenvalues of Aare simply $\lambda_1 = \alpha - 1$ and $\lambda_2 = \delta - 1$. It is straight forward to see that these eigenvalues satisfy the conditions of Theorem 1. Consequently, the zero solution of (13) is asymptotically stable, and the states of the controlled map (11) are asymptotically stabilized. □

The result of Theorem 2 can be easily put to the test. Consider, for instance, parameters $(\alpha, \beta, \gamma, \delta) = (0.9, -0.6013, 2, 0.5)$, initial conditions $(x(0), y(0)) = (-0.72, -0.64)$, and fractional order $v = 0.98$. Using a modified version of the numerical formula (8), we obtain the states depicted in Figure 8. Clearly, the states do converge towards the all-zero solution. Although the convergence was only established for the commensurate case, experiments have shown that the proposed control laws are also valid for the incommensurate case. Figure 9 shows the controlled states with the same parameters and initial conditions from above, but with different fractional orders $(v_1, v_2) = (0.99, 0.95)$. Again, we see that the states do in fact converge towards zero, indicating successful stabilization. However, it is apparent that the convergence happens faster in the commensurate case where $v_1 = v_2$.

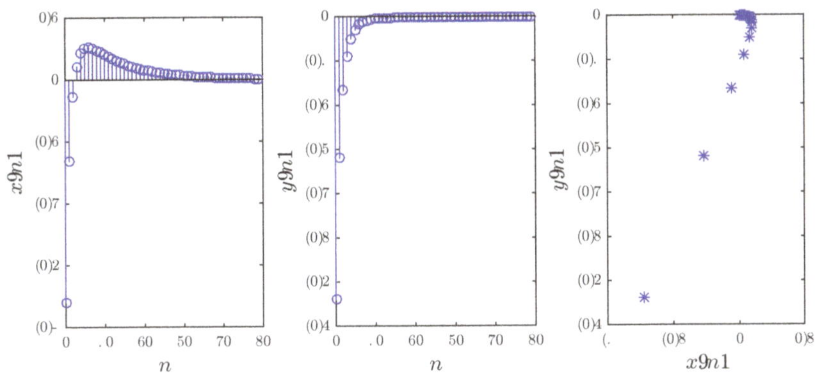

Figure 8. Stabilized states of the controlled fractional Tinkerbell map (11) with parameters $(\alpha, \beta, \gamma, \delta) = (0.9, -0.6013, 2, 0.5)$, initial conditions $(x(0), y(0)) = (-0.72, -0.64)$, and fractional order $v = 0.98$.

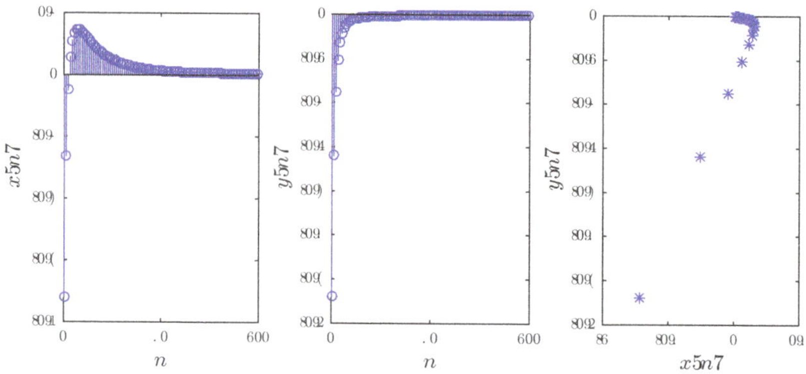

Figure 9. Stabilized states of the controlled fractional Tinkerbell map (11) with parameters $(\alpha, \beta, \gamma, \delta) = (0.9, -0.6013, 2, 0.5)$, initial conditions $(x(0), y(0)) = (-0.72, -0.64)$, and fractional orders $(v_1, v_2) = (0.99, 0.95)$.

5. Conclusions

In this paper, we have considered a fractional Caputo-difference form of the standard Tinkerbell chaotic map, which is well known for its rich dynamics and interesting characteristics. The dynamics of the fractional Tinkerbell map were investigated numerically using phase plots, bifurcation diagrams, and Lyapunov exponents. Through this investigation, we observed that the fractional order has a significant effect on the fractional map's dynamics. This confirms what has been reported previously in the literature and suggests that the fractional map is superior to the standard one, as it includes more degrees of freedom.

We have also introduced a feedback linearization stabilizing controller for the proposed map and established the asymptotic convergence of the states towards the all-zero solution by means of the stability theory of linear fractional discrete systems. The success of the proposed scheme was demonstrated through numerical simulations both in the commensurate and incommensurate cases.

Although feedback linearization is simple to design and implement, its practicality has been challenged by many in the control and cybernetics research communities. For future work, we plan to investigate other control schemes that can perform better in terms of the power consumption and other essential criteria. The main challenge that we anticipate is the limited literature concerning the stability of fractional discrete systems, especially on the Lyapunov method. This would have to be addressed in order to be able to establish the convergence of any new control scheme.

Author Contributions: Conceptualization, A.O.; investigation, A.-A.K.; methodology, A.O. and A.-A.K.; project administration, V.-T.P.; resources, S.B.; software, S.B. and T.P.V.; supervision, V.V.H.; validation, T.P.V.; writing, original draft, V.-T.P.; writing, review and editing, V.V.H.

Funding: This research received no external funding.

Conflicts of Interest: The authors declare no conflict of interest.

References

1. Lorenz, E.N. Deterministic nonperiodic flow. *J. Atmos. Sci.* **1963**, *20*, 130–141. [CrossRef]
2. Hénon, M. A two-dimensional mapping with a strange attractor. *Commun. Math. Phys.* **1976**, *50*, 69–77. [CrossRef]
3. May, R. Simple mathematical models with very complicated dynamics. *Nature* **1976**, *261*, 459–467. [CrossRef] [PubMed]
4. Lozi, R. Un atracteur étrange du type attracteur de Hénon. *J. Phys.* **1978**, *39*, 9–10.
5. Stefanski, K. Modelling chaos and hyperchaos with 3D maps. *Chaos Solitons Fractals* **1998**, *9*, 83–93. [CrossRef]

6. Itoh, M.; Yang, T.; Chua, L. Conditions for impulsive synchronization of chaotic and hyperchaotic systems. *Int. J. Bifurc. Chaos* **2001**, *11*, 551–558. [CrossRef]
7. Fiori, S. Nonlinear damped oscillators on Riemannian manifolds: Fundamentals. *J. Syst. Sci. Complex.* **2016**, *29*, 22–40. [CrossRef]
8. Fiori, S. Nonlinear damped oscillators on Riemannian manifolds: Numerical simulation. *Commun. Nonlinear Sci. Numer. Simul.* **2017**, *47*, 207–222. [CrossRef]
9. Huang, L.; Bae, Y. Chaotic dynamics of the fractional-love model with an external environment. *Entropy* **2018**, *20*, 53. [CrossRef]
10. Huang, L.; Bae, Y. Nonlinear behavior in fractional-order Romeo and Juliet's love model influenced by external force with fuzzy function. *Int. J. Fuzzy Syst.* **2018**, 1–9. [CrossRef]
11. Goodrich, C.; Peterson, A.C. *Discrete Fractional Calculus*; Springer: Berlin, Germany, 2015.
12. Wu, G.; Baleanu, D. Discrete fractional logistic map and its chaos. *Nonlinear Dyn.* **2013**, *75*, 283–287. [CrossRef]
13. Hu, T. Discrete chaos in fractional Hénon map. *Appl. Math.* **2014**, *5*, 2243–2248. [CrossRef]
14. Shukla, M.K.; Sharma, B.B. Investigation of chaos in fractional order generalized hyperchaotic Hénon map. *Int. J. Electron. Commun.* **2017**, *78*, 265–273. [CrossRef]
15. Khennaoui, A.; Ouannas, A.; Bendoukha, S.; Wang, X.; Pham, V.T. On Chaos in the Fractional–Order Discrete-Time Unified System and its Control Synchronization. *Entropy* **2018**, *20*, 530. [CrossRef]
16. Megherbi, O.; Hamiche, H.; Djennoune, S.; Bettayeb, M. A new contribution for the impulsive synchronization of fractional–order discrete-time chaotic systems. *Nonlinear Dyn.* **2017**, *90*, 1519–1533. [CrossRef]
17. Zhang, X.; Li, Z.; Chang, D. Dynamics, circuit implementation and synchronization of a new three-dimensional fractional-order chaotic system. *Int. J. Electron. Commun.* **2017**, *82*, 435–445. [CrossRef]
18. Khennaoui, A.; Ouannas, A.; Bendoukha, S.; Grassi, G.; Wang, X.; Pham, V. Generalized and inverse generalized synchronization of fractional-order discrete-time chaotic systems with non-identical orders and dimensions. *Adv. Differ. Equ.* **2018**, *2018*, 303. [CrossRef]
19. Bendoukha, S.; Ouannas, A.; Wang, X.; Khennaoui, A.; Pham, V.T.; Grassi, G.; Huynh, V. The co-existence of different synchronization types in fractional-order discrete-time chaotic systems with non-identical dimensions and orders. *Entropy* **2018**, *20*, 710. [CrossRef]
20. Ouannas, A.; Khennaoui, A.; Bendoukha, S.; Grassi, G. On the Q-S chaos synchronization of fractional-order discrete-time systems: general method and examples. *Discrete Dyn. Nat. Soc.* **2018**, *2018*, 2950357. [CrossRef]
21. Ismail, S.; Said, L.; Rezk, A.; Radwan, A.; Madian, A.; Abu-Elyazeed, M.; Soliman, A. Generalized fractional logistic map encryption system based on FPGA. *Int. J. Electron. Commun.* **2017**, *80*, 114–126. [CrossRef]
22. Aulbach, B.; Colonius, F. *Six Lectures on Dynamical Systems*; World Scientific: Singapore, 1996.
23. Nusse, H.; Yorke, J. *Dynamics: Numerical Explorations*; Springer: New York, NY, USA, 1997.
24. Davidchack, R.; Lai, Y.; Klebanoff, A.; Bollt, E. Towards complete detection of unstable periodic orbits in chaotic systems. *Phys. Lett. A* **2001**, *287*, 99–104. [CrossRef]
25. Mcsharry, P.; Ruffino, P. Asymptotic angular stability in non-linear systems: Rotation numbers and winding numbers. *Dyn. Syst.* **2003**, *18*, 191–200. [CrossRef]
26. Yuan, S.; Jiang, T.; Jing, Z. Bifurcation and chaos in the tinkerbell map. *Int. J. Bifurc. Chaos* **2011**, *21*, 3137–3156. [CrossRef]
27. Atici, F.M.; Eloe, P.W. Discrete fractional calculus with the nabla operator. *Electron. J. Qual. Theory Differ. Equ. Spec. Ed. I* **2009**, *3*, 1–12. [CrossRef]
28. Abdeljawad, T. On Riemann and Caputo fractional differences. *Comput. Math. Appl.* **2011**, *62*, 1602–1611. [CrossRef]
29. Anastassiou, G. Principles of delta fractional calculus on time scales and inequalities. *Math. Comput. Model.* **2010**, *52*, 556–566. [CrossRef]
30. Chen, F.; Luo, X.; Zhou, Y. Existence Results for Nonlinear Fractional Difference Equation. *Adv. Differ. Equ.* **2010**, *2011*, 713201. [CrossRef]

31. Cermak, J.; Gyori, I.; Nechvatal, L. On explicit stability condition for a linear fractional difference system. *Fract. Calc. Appl. Anal.* **2015**, *18*, 651–672. [CrossRef]
32. Wu, G.; Baleanu, D.; Luo, W. Lyapunov functions for Riemann–Liouville-like discrete fractional equations. *Appl. Math. Comput.* **2017**, *314*, 228–236.

applied sciences

MDPI

Article

Policy-Compliant Maximum Network Flows

Pieter Audenaert *, Didier Colle and Mario Pickavet

Faculty of Engineering and Architecture, Department of Information Technology, Ghent-University-imec, 9052 Ghent, Belgium; didier.colle@ugent.be (D.C.); mario.pickavet@ugent.be (M.P.)

* Correspondence: Pieter.Audenaert@UGent.be; Tel.: +32-(0)9-33-14900

Received: 19 January 2019; Accepted: 23 February 2019; Published: 28 February 2019

Abstract: Computer network administrators are often interested in the maximal bandwidth that can be achieved between two nodes in the network, or how many edges can fail before the network gets disconnected. Classic maximum flow algorithms that solve these problems are well-known. However, in practice, network policies are in effect, severely restricting the flow that can actually be set up. These policies are put into place to conform to service level agreements and optimize network throughput, and can have a large impact on the actual routing of the flows. In this work, we model the problem and define a series of progressively more complex conditions and algorithms that calculate increasingly tighter bounds on the policy-compliant maximum flow using regular expressions and finite state automata. To the best of our knowledge, this is the first time that specific conditions are deduced, which characterize how to calculate policy-compliant maximum flows using classic algorithms on an unmodified network.

Keywords: communication networks; maximum flow; network policies; algorithms

1. Introduction

Connecting two nodes in a computer communication network involves setting up paths, possibly more than one. Often, this is done with resiliency in mind: clearly, having more than one path between nodes, it might be possible to route around a failed link or node, depending on the paths themselves and the failed node or link. Also, network operators provide multiple paths between nodes in order to increase the maximum achievable throughput or flow in between these nodes.

In practice, however, network policies are in effect, effectively restricting the flow that can be set up between two nodes in the network. Specific restrictions can be implemented to optimize routing, due to security constraints or because of policies and agreements between different network operators. As one example, inter-domain paths in the internet often fulfill a valley-free routing constraint [1,2] which is a simple condition that models real-life business agreements between different operators. Compared to a policy-free routing model, these conditions severely restrict allowed paths and might have a large impact on the overall throughput between two nodes when the conditions are strictly enforced. Conditions also influence path diversity, which in turn has a large impact on the overall resiliency of the connection.

Calculating the maximum throughput that can be achieved between two nodes in a network can easily be done via classic techniques solving the maximum flow problem, such as the algorithm of Ford–Fulkerson or the push-relabel method [3,4]. However, these algorithms do not take into account policies and do not enforce any condition or constraint at all. In this work, we will adapt generic flow algorithms, such that the flow that is achieved actually is a policy-compliant maximum flow. Policies are defined using finite state automata (FSA) [5,6], as their expressive power is sufficient for many purposes in real life, while still being conceptually simple.

We will define an algorithm that exactly solves the policy-compliant maximum flow problem, however, at the expense of a large computational footprint. Then, we will continue to adapt the classic algorithms and datastructures such that they take into account policies that can be expressed using FSA. These algorithms, however, do not guarantee exact solutions to the problem anymore; they merely provide lower bounds to the exact answer. At each step, we are able to tighten the lower bound, such that it comes closer to the exact solution. However, each step also implies additional computational work, so we are trading off the quality of the results with the time needed to run the algorithms.

The rest of the paper is organized as follows. First, in Section 2 we provide some background about policies and how they can model certain conditions and constraints that are applied in real life. We also point out some difficulties that arise when trying to adapt the classic algorithms. Then, in Section 3 we discuss the formal model and the classic algorithms from literature, and how we can formally model policy-compliant connections. Next, in Section 4 we adapt the algorithms that were discussed in Section 3 to obtain a series of techniques that allow us to calculate lower bounds to the problem. Finally, Section 5 discusses the results obtained, and concludes this work, after which we provide an outlook to future work.

2. Background and Context

Policies are very important in today's internet. Specific agreements between operators are often hidden from prying eyes, but they can have a huge impact on how specific paths are set up between them. It turns out that many network policies, used in practice, can be translated in very simple formal languages, all belonging to a class of formal languages called "regular languages" [5,6]. This class is one of the most basic classes of languages used in computer science, as it is non-trivial but contains a very broad range of languages with widely differing properties that are very useful in practice.

The complexity of properties which we can express using regular languages is larger than would be expected at first sight. To provide a non-trivial, albeit contrived example, consider a government that wants to contact its embassy located in another country. Some links are trusted, as they are operated by friendly states, but some links in between may be eavesdropped upon by a rival state. Traffic crossing such links needs to be encrypted; once encrypted, traffic may be routed through any link in the network. The receiving party will need to decrypt the message, before it can be distributed to internal data-processing departments. The government thus places into force the following conditions: traffic from the government to the embassy should either be sent through links owned exclusively by friendly states, or it should pass the encryption box, after which it can be routed through any link. However, if encrypted, the message has to be decrypted before delivery at the internal departments at the embassy, i.e., pass through a specific decryption box.

It is clear that enforcing such policies is important, moreover such policies should be formalized instead of specified in vague terms as above. Regular languages fit that job quite well, as they allow policies like the one above, next to other much more involved policies, to be specified without any possibility for misunderstanding. Moreover, once specified, network-routes can be automatically validated against a set of policies, and compliance can be trivially checked.

Taking into account policies in the network, a natural question is to ask for their impact on the routing between nodes in the network. More specifically, in this work we are mainly interested in their impact on the maximum flow achievable between two nodes. Indeed, previously acceptable paths suddenly do not obey the specific policies, and as such, some routes, maybe all, will become invalid. Thus, the policy-compliant maximum flow between two nodes needs to be redefined and recalculated. Gaining insight into the flow-structure between two nodes also tells us something about resiliency in the network. Operators might be convinced their network has sufficient spare capacity, however, depending on the policy at hand and the specific structure of the network, a failing link might have tremendous consequences as re-routing traffic around the failure might invalidate the policy which is unacceptable.

We thus come to the following main problem-statement that will be treated in this work: how to calculate or approximate the policy-compliant maximum flow between two nodes?

Clearly, policy-compliance and both the theoretical and practical consequences are important in the design and operation of networks. From a practical point of view, much information can be found about policies and being compliant, and multiple frameworks for the management and monitoring of policies exist and are in current use. In contrast, few theoretical models have been developed, and literature is sparse when taking a more fundamental approach to the problem.

Caesar and Rexford [7] discuss how routing policies came into existence, and how the protocols have evolved over time and became increasingly complex. They discuss how common policies are implemented and address the problems that arise in applying and supporting policies. Feamster et al. [8] discuss fundamental objectives for interdomain routing and traffic engineering and provide practical guidelines. They show how greater flexibility can be gained in several situations and demonstrate the manipulation of traffic via small changes in specific routes. The paper by Hu et al. [9] proposes an approach to overcome the inherent constraints of compliant recovery schemes. Adapting protocols, they succeed in improving route diversity, in turn increasing resilience.

Klöti et al. [10] proposed a graph-transformation technique that constructs a tensor product of a graph and a finite state automaton. They show how to model policies using FSA and apply standard flow maximization algorithms on the transformed graph in order to obtain bounds for the policy-compliant maximum flow problem. Sobrinho et al. [11] provide a deep mathematical analysis to gain insight in the inner workings of route-vector protocols. They relate this to a class of routing policies and quantify how much intrinsic connectivity is lost due to typical routing policies. Erlebach et al. [12] approximate the maximum number of edge- and node-disjoint valid paths between two nodes, via an involved mathematical theory and specific approximation-algorithms.

Soulé et al. [13] provide a declarative formal language based on logical formulas and regular expressions to express network policies in the Merlin framework. After compilation, a constraint solver allocates paths. Tools are provided to verify whether conditions are violated. Hinrichs et al. [14] introduce a declarative policy management language. They focus on expressive power so that policies can be expressed naturally while still being able to enforce policies efficiently. They focus on enterprise networks, and their design using formal mathematical languages is attractive.

Raghavan et al. [15] provide a practical authenticated source routing system that allows for fine-grained path selection and enforcement of cryptographic policies. Capabilities can be defined and composed resulting in complex statements defining specific policies.

Godfrey et al. [16] introduce so-called pathlet routing, pathlets being building blocks which can be concatenated to obtain end-to-end routes fulfilling policy constraints. Pathlet routing can handle typical policies, but also other recent multipath routing proposals or source-routing approaches. Batista et al. [17] propose a policy-based OpenFlow network management framework using the Ponder policy definition language. They focus on simplicity of the system, trying to infer both static and dynamic conflicts. In [18] they update their work and propose a more theoretical approach to conflict detection using first-order mathematical logic to model flows. A Prolog rule-engine applies condition-action rules to infer problems. Xu and Rexford [19] discuss problems concerning multipath interdomain routing and introduce MIRO (Multi-path Interdomain ROuting), which gives transit domains control over the flow across their infrastructure. MIRO remains backwards compatible with other technologies and offers large flexibility with reasonable overhead.

3. Graph Model and Basic Algorithms

We now introduce basic concepts that are needed in this work. Intuitively, we can think about a flow network like an oil-producing plant, where oil is drilled at one site and needs to be transported to an oil depot via a set of pipes. Following the flow from the source to the sink, the oil can be split or joined at junctions. Most important here is the fact that the path that was followed by a certain molecule of oil is of no importance: it carries no history and passes through junctions anonymously. In contrast, in this work, we need to pay attention to the specific path that a certain amount of flow has followed: it carries its history with itself, see Figure 1. The amount of history that accompanies the flow is, however, limited: it is represented via its state, and the number of possible different states is limited by the size of the automaton (cf. infra). To understand how to model policies using FSA, we refer to [10].

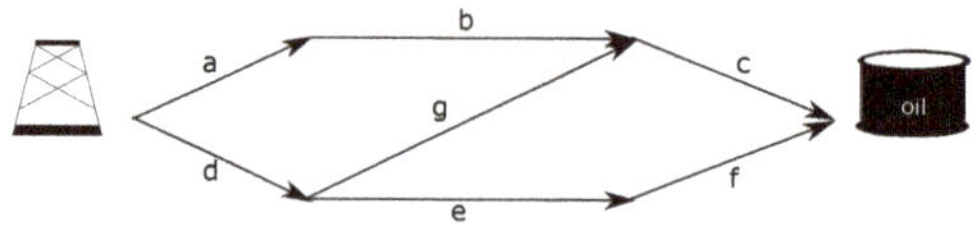

Figure 1. It doesn't matter for oil whether it followed path abc, path def or path dgc, but in this paper, we do care about the specific paths followed. Policies about which path is acceptable and which path is not, are expressed via constraints on the labels across the paths, i.e., abc, def and dgc.

3.1. Flow Networks and Flows

We now introduce flow networks and flows, following the approach taken by Cormen et al. [4]. These definitions might differ from other approaches in several aspects, e.g., the fact that the capacity function c and the flow f are total functions, allowing more concise and elegant definitions of constraints. The rationale behind this way of defining the context is included in Cormen et al. [4] and we refer interested readers to study the mathematical discussions in that book. We thus define a flow network $G = (V, E)$ as a directed graph together with a capacity function $c : V \times V \to \mathbb{R}$ such that each edge $(u, v) \in E$ has a nonnegative capacity $c(u, v) \geq 0$. Note that c is a total function such that, if $(u, v) \notin E$, we have $c(u, v) = 0$. Moreover, we require that if E contains a certain edge (u, v), then there is no reverse edge (v, u) present in E. We also disallow loops such that $\forall u : (u, u) \notin E$. Two specific nodes in the graph are now chosen out of V, one vertex being the source s and one vertex being the sink t. We now want to send a flow from s to t, a flow being a total function $f : V \times V \to \mathbb{R}$ subject to the following constraints:

- A capacity constraint which expresses that flow can never exceed capacity:

$$\forall u, v \in V : 0 \leq f(u, v) \leq c(u, v) \tag{1}$$

- A flow conservation constraint which expresses that for all nodes (except source and sink) the incoming flow equals the outgoing flow:

$$\forall u \in V - \{s, t\} : \sum_{v \in V} f(v, u) = \sum_{v \in V} f(u, v) \tag{2}$$

Note that $f(u, v) = 0$ if $(u, v) \notin E$.

The value $|f|$ of a flow is defined as the flow leaving the source minus the flow arriving at the source: $|f| = \sum_{v \in V} f(s, v) - \sum_{v \in V} f(v, s)$. Equivalently, we can define the value of the flow just as well as $|f| = \sum_{v \in V} f(v, t) - \sum_{v \in V} f(t, v)$. Note that one can define the value of the flow using a formula containing only s but not t, or a formula containing only t but not s; one never needs both s and t in one formula at the same time, see the book by Cormen et al. [4]. Also note that we allow $\sum_{v \in V} f(v, s)$ as well as $\sum_{v \in V} f(t, v)$ to be strictly larger than 0. The classic problem, given a flow network G, is to maximize the flow $|f|$ from s to t under the constraints above. (The $|.|$-notation denotes flow value, not absolute value or cardinality.)

3.2. Residual Graphs and Augmenting Paths

One line of solving this so-called maximum flow problem is via the introduction of residual graphs and the calculation of augmenting paths. In short, we iteratively increase the value of the flow, starting with $\forall u, v \in V : f(u, v) = 0$. At each step, we calculate an augmenting path in the residual graph, which is nothing but the original graph, extended with return-edges. We will now formally define the necessary concepts.

First, we define the residual capacity $c_f : V \times V \to \mathbb{R}$ in the following way. We iterate over all edges $(u,v) \in E$ and for each such edge we define the following:

$$\begin{cases} c_f(u,v) = c(u,v) - f(u,v) \\ c_f(v,u) = f(u,v) \end{cases} \tag{3}$$

If there is no edge between u and v in any direction, i.e., $(u,v) \notin E$ and $(v,u) \notin E$, then both $c_f(u,v) = 0$ and $c_f(v,u) = 0$.

Next, we define the residual network $G_f = (V, E_f)$, where $E_f = \{(u,v) \in V \times V : c_f(u,v) > 0\}$. So, for each edge in the original graph we have a residual edge if the capacity of that edge is not filled up yet, and moreover we add a reverse edge when it carries any flow at all.

This residual graph will now be used to find augmenting paths. Given a flow network $G = (V, E)$ and a flow f, an augmenting path p is defined as a path from s to t in the residual network, that is $p = [e_1, e_2, \ldots, e_n]$, such that

- edges belong to E_f, thus
$$\forall e_i : e_i \in E_f, \tag{4}$$
- we start in s, thus
$$\exists v : e_1 = (s, v), \tag{5}$$
- succeeding edges are connected, thus
$$\forall i \in \{1, \ldots, n-1\} : \exists u, v, w : e_i = (u,v) \land e_{i+1} = (v,w) \tag{6}$$
- we end in t, thus
$$\exists v : e_n = (v, t). \tag{7}$$

In order to increase the flow f, we will change the flow along the edges of the augmenting path p with the maximum amount possible, namely the residual capacity of the path p which is

$$c_f(p) = min\{c_f(u,v) : (u,v) \in p\}. \tag{8}$$

The flow f can now be increased to a new flow f' as follows:

$$f'(u,v) = \begin{cases} f(u,v) + c_f(p) & \text{if } (u,v) \in p, \\ f(u,v) - c_f(p) & \text{if } (v,u) \in p, \\ f(u,v) & \text{otherwise.} \end{cases} \tag{9}$$

Iteratively finding augmenting paths in a residual graph and increasing the flow will finally come to a stop when there are no augmenting paths anymore. Then, the maximum flow has been reached. Depending on the procedure used to find augmenting paths, the above algorithm is called Ford–Fulkerson or Edmonds–Karp.

3.3. Realization of a Flow

Apart from calculating the value of the maximum flow in a flow network, we are also interested in the actual paths followed by the flow; this set of paths is called the "realization" of the flow. Augmenting paths are a device to calculate the maximum flow, however, augmenting paths are not real paths in the flow network as they might comprise edges from E_f that are not in E. Define a path p

from s to t in G in exactly the same way as an augmenting path, except that edges in the path should be contained in E instead of E_f. Thus, in addition to (5)–(7), we replace (4) with

$$\forall e_i : e_i \in E. \tag{10}$$

Given a flow network G and accompanying maximum flow f, it is possible to determine a set

$$S = \{(p_1, c_1), (p_2, c_2), \ldots, (p_n, c_n)\} \tag{11}$$

that realizes that flow. Intuitively, we start with no flow in the network at all. Then, for each of the paths p_i we fill all edges of that path with an additional amount c_i. After all paths are processed, we obtain the maximum flow f.

Construction of this set S can be done simultaneously with the calculation of the maximum flow. During execution of the maximum flow algorithm, we perform the following steps every time a new augmenting path p with additional flow amount c has been found, starting with an empty set $S = \{\}$.

1. We split the augmenting path in forward sections and backward sections. That is, if $p = [e_1, e_2, \ldots, e_n]$, we cut it down and obtain $[[e_1, \ldots, e_i], [e_j, \ldots, e_k], \ldots, [e_l, \ldots, e_n]]$ such that every section $[e_x, \ldots, e_y]$ contains only edges from E or only edges from $E_f - E$. Without losing generality, we only discuss the case where the augmenting path p consists of two forward sections with one backward section in between, thus $p = [[e_1, \ldots, e_i], [e_j, \ldots, e_k], [e_l, \ldots, e_n]]$ such that $\{e_1, \ldots, e_i, e_l, \ldots, e_n\} \subseteq E$ and $\{e_j, \ldots, e_k\} \subseteq E_f - E$.
2. For this backward section $[e_j, \ldots, e_k]$ there must exist a path p' carrying some flow in the opposite direction, which the augmenting path p is canceling out. Indeed, as $[e_j, \ldots, e_k]$ are edges in E_f, they have been introduced in the residual graph G_f thanks to the fact that the corresponding forward edges in E have already been used in a previously found path. Thus, if $e_j = (u_j, v_j)$ and $e_k = (u_k, v_k)$, then S will already contain an item (p', c') where the path $p' = [\ldots, (v_x, v_k), (v_k, u_k), \ldots, (v_j, u_j), (u_j, u_y), \ldots]$, as the augmenting path p is crossing that section in the reverse direction, see Figure 2.
3. Remove this item (p', c') from S, and add two new paths $[e_1, \ldots, e_i, (u_j, u_y), \ldots]$ and $[\ldots, (v_x, v_k), e_l, \ldots, e_n]$, each with amount c. As such, the augmenting path has been cut and glued together with a previously found path, resulting in two new paths.
4. Now, it might be that $c < c'$. In that case, path p' is not to be removed from S entirely, as it still carries some flow that was not canceled by p. Add $(p', c' - c)$ to S.

Thus, each iteration of the maximum flow algorithm results in an augmenting path that is possibly cut in sections to remove reversed edges from the residual graph. Forward sections are glued together with sections from previously found paths. When no augmenting paths can be found, the set S will contain paths and associated flow amounts, that together sum up to the maximum flow. That is, if upon termination $S = \{(p_1, c_1), \ldots, (p_n, c_n)\}$, we have

$$f(u, v) = \sum_{\{(p_i, c_i) : (u,v) \in p_i\}} c_i \tag{12}$$

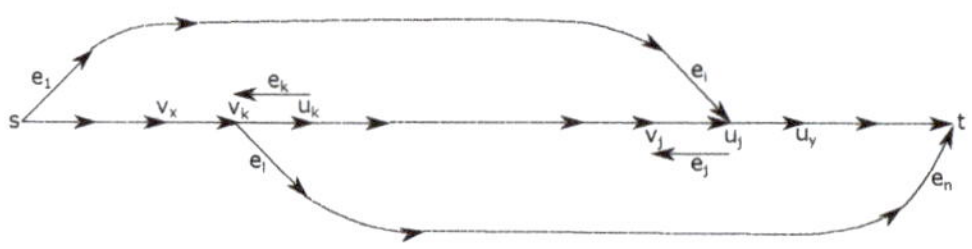

Figure 2. Three sections: forward, backward and forward.

Thanks to the fact that every flow has a realization and the fact that it is easy to derive the flow from a realization, from now on we identify a realization of a flow and the flow itself, and use the two concepts as one.

4. Policy-Compliant Paths and Algorithms

We now continue to define the concept of policy-compliant paths. Then we want to find the maximum flow in the network, allowing only the use of such policy-compliant paths. We will provide algorithms that calculate lower bounds, as well as exact solutions.

4.1. Finite State Automata

We now proceed to define finite state automata [5,6], which are the simplest of all classic automata. Note that we will only consider deterministic automata, although for finite state automata we have that non-determinism does not add expressive power, i.e., the class of deterministic finite state automata is exactly the same as the class of non-deterministic finite state automata.

We define a deterministic finite state automaton as $M = (Q, \Sigma, \delta, q_0, F)$, where

- Q is a finite set of internal states,
- Σ is a finite set of symbols, called the input alphabet,
- $\delta : Q \times \Sigma \to Q$ is the (total) transition function,
- $q_0 \in Q$ is the start (or initial) state,
- $F \subseteq Q$ is the set of final states.

Such an automaton works as follows. Initially, we start with an input word $w \in \Sigma^*$, where Σ^* is the set of strings obtained by concatenating zero or more symbols from Σ. Each of these symbols of the word w gets consumed, from left to right, while the automaton changes its state accordingly to its transition function, taking as input the current state and the symbol just read. If the automaton starts in state q_0, and finishes reading the word w while reaching state q, then we say that the word w is accepted by the automaton if and only if $q \in F$. For convenience, we introduce the extended transition function $\delta^* : Q \times \Sigma^* \to Q$, where the second argument is a string rather than a single symbol. We define δ^* recursively as follows:

$$\delta^*(q, w) = \begin{cases} q & \text{if } w = \lambda, \\ \delta^*(\delta(q, a), v) & \text{if } w = av, \end{cases} \tag{13}$$

where $a \in \Sigma$ and $v, w \in \Sigma^*$. Note that we also introduced the empty word λ out of convenience for this definition. We won't need it any further in this work.

4.2. Labeled Networks and Policy-Compliant Paths

Given a certain flow network $G = (V, E)$ and an FSA M, we define a labeling function $l : E \to \Sigma$ that attaches one symbol to every edge of the graph, which becomes a labeled flow network. Before, we were interested in all paths from source to sink, in order to maximize the total flow value of the network. Now, we constrain the allowed paths to only those that are accepted by the FSA M. Formally, we define a policy-compliant path p from s to t in G to be a sequence of edges from E, such that the symbols that are encountered, in order, constitute a word that is accepted by M. Thus, in addition to (5)–(7) and (10), we add the following constraint:

$$\delta^*(q_0, l(e_1)l(e_2)\dots l(e_n)) \in F \tag{14}$$

Thus, we are looking for paths from source to sink that are accepted by the automaton. Now we can define the policy-compliant maximum flow problem (PCMF) as calculating the maximum flow $|f|$

from s to t in a labelled flow network G_f, subject to the constraint that all paths that realize the flow f are accepted by an FSA M, i.e., they are policy-compliant paths.

4.3. Brute Force Approach

A basic, inefficient way of solving the policy-compliant maximum flow problem is by way of brute force. In its essence, we first calculate all policy-compliant paths from source to sink, and afterward try out all different orders in which these paths can be filled up. The maximum flow will be reached by at least one of these orderings. As such, this approach will result in the exact solution, however, at the expense of large computational complexity and memory requirements. Indeed, as we employ our search via e.g., a depth-first-search, it might be the case that we need to visit a certain vertex $u \in E$ more than once, because the internal state of the FSA might be different. Effectively, we need to build a new graph G_{BF} that contains tuples of vertices and states. After appropriately defining the edges between these tuples, a typical depth-first-search might be used to enumerate successively all policy-compliant paths. More formally, we perform the following steps.

1. Define G_{BF} as the graph with $V_{BF} = \{(u,q) : u \in V \wedge q \in Q\}$ and $E_{BF} = \{((u_1,q_1),(u_2,q_2)) : u_1, u_2 \in V \wedge q_1, q_2 \in Q \wedge (u_1,u_2) \in E \wedge \delta(q_1, l((u_1,u_2))) = q_2\}$. That is, the vertices of G_{BF} are all possible tuples of vertices in V and states in Q, and the vertices in G_{BF} are connected if and only if the corresponding vertices u_1 and u_2 in V were connected by an edge $e \in E$, such that the FSA changes its state from q_1 to q_2 when reading the symbol $l(e)$ associated with the original edge. Thus, we are defining G_{BF} as the Cartesian product of V and Q.
2. Start a depth-first-search from (s,q_0) to all (t,q) such that $q \in F$. This will produce all paths of the form $p = [e_1,\ldots,e_n]$ with $\forall e_i : e_i \in E_{BF}$ while $\exists v \in V_{BF} : e_1 = ((s,q_0),v)$ and $\exists v \in V_{BF} : \exists q \in F : e_n = (v,(t,q))$. Every such path corresponds to a policy-compliant path in G, and every policy-compliant path in G will be found this way. Collect all paths found by this depth-first-search in the set $BF = \{p_1,\ldots,p_n\}$.
3. Generate all permutations of the elements in this set BF; this results in $n!$ sequences of the n paths. Each of these sequences will correspond to an ordering in which we fill the paths to their residual capacity at that time. Note that it is not necessary to generate all permutations at once. It is sufficient to do this iteratively via e.g., the Steinhaus–Johnson–Trotter algorithm.
4. Each such sequence $s = [p_1^s,\ldots,p_n^s]$ contains exactly the same items as BF itself. We define an empty flow $f_s(u,v) = 0$ and iterate through this sequence: for each path $p_i^s = [e_{i,1}^s,\ldots,e_{i,m}^s]$ we calculate $min_i^s = min\{c(e_{i,j}^s) - f_s(e_{i,j}^s) : 1 \leq j \leq m\}$ which represents the largest amount of flow that can be accommodated along the path p_i^s at that moment. The flow f_s can now be increased to a new flow f_s' as follows:

$$f_s'(u,v) = \begin{cases} f_s(u,v) + min_i^s & \text{if } (u,v) \in p_i^s, \\ f_s(u,v) & \text{otherwise.} \end{cases} \quad (15)$$

5. After all paths p_i^s are processed, the final flow f_s will be a lower bound for the policy-compliant maximum flow for the network G. Thus, $max\{|f_s|\}$ will be the value of the policy-compliant maximum flow.

It is clear that this approach results in the exact value of the solution to the policy-compliant maximum flow problem. However, except for small networks and automata it will quickly take too much space and time to execute.

4.4. First Lower Bound

To cut down on the needed computational resources, we start from the classic Ford–Fulkerson or Edmonds–Karp algorithm. In its essence, these algorithms start with an empty flow, after which a new augmenting path is looked for in the residual graph. If found, this augmenting path gives rise to an augmented flow, after which the residual graph is updated. Then, a new iteration is started. Once no new augmented paths can be found, the algorithm has reached the maximum flow.

In order to apply this approach to the policy-compliant maximum flow problem, it will be necessary to keep track of how the flow actually is realized, throughout the execution of the algorithm. This relates to the fact that flows cannot merge and split anonymously anymore, as discussed previously.

As a first step, it will be necessary to keep track of the different state-transitions that occur when flow crosses an edge $e = (u, v) \in E$ in a path p and changes state during that transition from state $q_u \in Q$ in vertex u to state $q_v \in Q$ in vertex v. To that aim, we define transitions t to be members of the set $Q \times Q = \{(q_u, q_v) : q_u \in Q \wedge q_v \in Q\}$. For every edge $e \in E$ we will now keep track of all transitions that cross that edge, and thus define the total function $T : V \times V \times Q \times Q \to \mathbb{R}$. This function keeps track of the amount of flow that crosses the edge (u, v) while changing state from q_u to q_v as the value $T(u, v, q_u, q_v)$. As a direct consequence, we have at all times $\forall u, v \in V$: $\sum_{q_u, q_v \in Q} T(u, v, q_u, q_v) = f(u, v) \leq c(u, v)$ which expresses the fact that a flow over an edge can be split up in different transitions, the amount of which add up exactly to the amount of flow crossing that very edge.

Keeping track of this function T allows the algorithm to discriminate between the different states in one single vertex, and the amount of the flow that is actually passing through that vertex having that state at that moment. This information is then used to glue together already found policy-compliant paths and a newly found augmenting path. In detail, this procedure works as follows.

1. Beginning with a labeled network G and a FSA M, we start a new flow and set $f(u, v) = 0$ for all vertices u and v. Moreover, we set $T(u, v, q_u, q_v) = 0$ for all vertices $u, v \in V$ and all states $q_u, q_v \in Q$.
2. Finding the first policy-compliant augmenting path can be straightforwardly done. Indeed, as the residual graph G_f is still the same as the original graph G (because $\forall u, v : f(u, v) = 0$), the augmenting path that is found in the first iteration will be a path containing only forward sections. The flow is then updated according to (9).
3. From the second iteration on, we need to take into account the reverse edges in the residual graph with great care. Now, it is not sufficient anymore to simply look for policy-compliant paths containing only forward sections, as an augmenting path may now contain reverse edges from the residual graph. In general, we will look for an augmenting path $p = [e_1, \ldots, e_n]$ where $e_i \in E_f$, that can have multiple forward and backward sections, and which can thus be split as $p = [[e_1, \ldots, e_i], [e_j, \ldots, e_k], \ldots, [e_l, \ldots, e_n]]$, such that each section $[e_x, \ldots, e_y]$ only contains edges from either E or $E_f - E$. We will now determine the constraints that are to be applied to ensure that the newly found augmenting path p can be split up and glued together with other policy-compliant paths that are already part of the flow f.
4. For simplicity, we discuss the case where the augmenting path p consists of two forward sections with one backward section in between, that is $p = [[e_1, \ldots, e_i], [e_j, \ldots, e_k], [e_l, \ldots, e_n]]$ such that $\{e_1, \ldots, e_i, e_l, \ldots, e_n\} \subseteq E$ and $\{e_j, \ldots, e_k\} \subseteq E_f - E$. To identify the vertices involved, suppose that $e_j = (u_j, v_j)$ and $e_k = (u_k, v_k)$, so $p = [[e_1, \ldots, e_i], [(u_j, v_j), \ldots, (u_k, v_k)], [e_l, \ldots, e_n]]$, see Figure 2. Starting from state q_0, the start state of the automaton, we will reach state $q_{u_j} = \delta^*(q_0, l(e_1) \ldots l(e_i))$ after we crossed all edges $[e_1, \ldots, e_i]$ in the first forward section of the augmenting path.
5. The first forward section of the augmenting path that we have found will be cut at this point, at the cut-vertex u_j. For this first approach to obtain a lower bound for the PCMF problem, we assert the overly restrictive constraint that $\exists u_y \in V$ and $\exists q_{u_y} \in Q$ such that

$$(u_j, u_y) \in E \wedge T(u_j, u_y, q_{u_j}, q_{u_y}) > 0. \tag{16}$$

Choose such an edge $e = (u_j, u_y) \in E$ and state $q_{u_y} \in Q$. When the augmenting path p is cut at vertex u_j, we can glue the new flow that comes in via the forward section $[e_1, \ldots, e_i]$ to the flow that leaves vertex u_j in state q_{u_j} via edge $e = (u_j, u_y)$ and reaches state q_{u_y} at the other end in vertex u_y. This flow is already available and does not concern us anymore.

6. Now, we need to cancel flow that used to cross the edges $[(v_k, u_k), \ldots, (v_j, u_j)]$, that is the backward section of the augmenting path p, crossed in the opposite direction. This gives rise to a second condition for this algorithm to find augmenting paths. This section contains a flow, reaching state q_{u_j} at the end. Tracing back where this flow comes from, we require that there must exist some flow passing vertex v_k in state $q_{v_k} \in Q$ which crosses the edges $[(v_k, u_k), \ldots, (v_j, u_j)]$ finally reaching vertex u_j in state q_{u_j}. This observation immediately translates to the following condition:

$$\exists q_{v_k} \in Q : \delta^*(q_{v_k}, l((v_k, u_k)) \ldots l((v_j, u_j))) = q_{u_j}. \tag{17}$$

Choose such a $q_{v_k} \in Q$. Then, the flow that is already present in the network and passes vertex v_k in state q_{v_k} will be canceled over the section $[(v_k, u_k), \ldots, (v_j, u_j)]$.

7. We now need to consider the last forward section $[e_l, \ldots, e_n]$. This section starts in vertex v_k and ends in the sink t, thus $\exists v \in V : e_l = (v_k, v)$ and $\exists v \in V : e_n = (v, t)$. We know that, having reached state q_{v_k} in v_k, we will cross edges $[e_l, \ldots, e_n]$ and need to reach some state $q_t \in F$ to obtain a policy-compliant augmenting path. This immediately gives rise to the third and last condition, which can be stated as

$$\delta^*(q_{v_k}, l(e_l) \ldots l(e_n)) = q_t \in F. \tag{18}$$

8. Suppose all conditions (16)–(18) for this augmenting path are fulfilled, which means that we have found an augmenting path that can be joined with the already available policy-compliant paths to increase the total policy-compliant flow. We now need to calculate the amount of flow that can be sent over this augmenting path. First, define $c_{u_j} = min\{c_f(u,v) : (u,v) \in [e_1, \ldots, e_i]\}$ to be the capacity that is at most available along the first forward section of the augmenting path and thus at the endpoint u_j of this section. Likewise, define $c_{v_k} = min\{c_f(u,v) : (u,v) \in [e_l, \ldots, e_n]\}$ to be the amount of flow that is at most available along the second forward section of the augmenting path, and thus the maximum amount of flow that can be diverted at the starting point v_k of this section. For the backward section of the augmenting path, define $c_{v_k u_k} = T(v_k, u_k, q_{v_k}, q_{u_k}), \ldots, c_{v_j u_j} = T(v_j, u_j, q_{v_j}, q_{u_j})$, to determine the already crossing flow over the backward edges, taking into account the state-transitions which we derived above. Taking everything together, we have

$$c_f(p) = min\{c_{u_j}, c_{v_k u_k}, \ldots, c_{v_j u_j}, c_{v_k}\} \tag{19}$$

(cf. (8)). This denotes the additional flow $c_f(p)$, under the assumption that $c_f(p) > 0$, that can now be sent over the newly found augmenting path p, such that the resulting flow f can be realized via only policy-compliant paths.

9. We now want to update the value of the flow f. This can be done via (9). However, we also need to update the specific values of the function T, for all edges involved in the augmenting path. First, we update T for all edges $e = (u_x, u_y)$ in the forward section $[e_1, \ldots, e_i]$ via $T'(u_x, u_y, q_{u_x}, q_{u_y}) = T(u_x, u_y, q_{u_x}, q_{u_y}) + c_f(p)$. Thus, for such an edge $e = (u_x, u_y)$ we add an additional $c_f(p)$ units of flow that change state from q_{u_x} to q_{u_y} while crossing edge e. For all edges $[(v_k, u_k), \ldots, (v_j, u_j)]$, which are crossed in the reverse direction in the backward section of the augmenting path, we set $T'(u_x, u_y, q_{u_x}, q_{u_y}) = T(u_x, u_y, q_{u_x}, q_{u_y}) - c_f(p)$, actually canceling out the pre-existent transitions in that section. Finally, for the edges in the final forward section $[e_l, \ldots, e_n]$ we update the transitions via $T'(u_x, u_y, q_{u_x}, q_{u_y}) = T(u_x, u_y, q_{u_x}, q_{u_y}) + c_f(p)$. This concludes the updates of

the transitions-function T and the work that needs to be done when an augmenting path has been found.

10. Iterate steps (3)–(9), that is, keep searching for augmenting paths that fulfill conditions (16)–(18) and update the flow f according to (9) and the transition-function T via the equations above. Once no more augmenting paths are found, stop iterating.
11. The resulting flow f can be realized via a set $\{p_1, \ldots, p_n\}$ of n policy-compliant paths p_i. Thus, the value $|f|$ is a lower bound for the value of the policy-compliant maximum flow problem.

4.5. Second Lower Bound

Although the approach described above results in correct lower bounds for the policy-compliant maximum flow problem, we point out that the conditions (16)–(18) that are imposed upon augmenting paths are very restrictive. Indeed, it is easy to describe specific cases where the approach misses additional augmenting paths due to these conditions, which can easily be relaxed at the expense of additional computational work. The next approach, which will result in a second lower bound that is tighter than the first bound, starts by relaxing condition (16).

In its essence, this condition says that for an augmenting path, containing a backward section, it is necessary to glue the immediately previous forward section to a pre-existent policy-compliant path at the cut-vertex, via an exact match of the state in that cut-vertex. Clearly, this is too restrictive a condition as it would be sufficient if the augmenting flow arriving at that vertex could be continued via the same edges, however taking different intermediate states in between.

1. To make this formal, consider the augmenting path $p = [e_1, \ldots, e_n]$ where $e_i \in E_f$, which may have multiple forward and backward sections, and which can thus be split as $p = [[e_1, \ldots, e_i], [e_j, \ldots, e_k], \ldots, [e_l, \ldots, e_n]]$, such that each section $[e_x, \ldots, e_y]$ only contains edges from either E or $E_f - E$. For simplicity, we once more discuss the case where the augmenting path p consists of two forward sections with one backward section in between, that is $p = [[e_1, \ldots, e_i], [e_j, \ldots, e_k], [e_l, \ldots, e_n]]$ such that $\{e_1, \ldots, e_i, e_l, \ldots, e_n\} \subseteq E$ and $\{e_j, \ldots, e_k\} \subseteq E_f$. To identify the vertices involved, say that $e_j = (u_j, v_j)$ and $e_k = (u_k, v_k)$, so $p = [[e_1, \ldots, e_i], [(u_j, v_j), \ldots, (u_k, v_k)], [e_l, \ldots, e_n]]$. The state at u_j is defined as $q_{u_j} = \delta^*(q_0, l(e_1) \ldots l(e_i))$.
2. For flow f, suppose that the policy-compliant paths $P = \{p_1, \ldots, p_n\}$ are the part of the realization of f that crosses u_j. Thus, $\forall p_i \in P : \exists v_i \in V : (u_j, v_i) \in p_i$. For each of these policy-compliant paths p_i, the state upon arriving in u_j is denoted by $q^i_{u_j}$. Moreover, there might be more than one path $p_i \in P$ that arrives in u_j with state $q^i_{u_j}$, that is for some i and j it might be the case that $i \neq j$ and $q^i_{u_j} = q^j_{u_j}$. The amount of flow that policy-compliant path p_i carries is denoted by c_i, thus $\forall p_i \in P : (p_i, c_i) \in S$, cf. (11).
3. Now, instead of stating in the first condition (16) that there must be a policy-compliant path $p_i \in P$ such that $q_{u_j} = q^i_{u_j}$ with $c_i > 0$ it is, in fact, sufficient that there exists a path $p_i \in P$ such that following the rest of the policy-compliant path p_i after crossing u_j results in a final state at the sink t even though $q_{u_j} \neq q^i_{u_j}$. More formally, if $p_i = [e^i_1, \ldots, e^i_k, e^i_l, \ldots, e^i_n]$ such that $e^i_k = (u_x, u_j)$ and $e^i_l = (u_j, u_y)$ for some $u_x \in V$ and $u_y \in V$, then it is sufficient to impose, instead of (16), simply that
$$\exists p_i \in P : \delta^*(q_{u_j}, l(e^i_l) \ldots l(e^i_n)) \in F. \tag{20}$$
Choose such a $p_i \in P$. That way, we are confident that the prefix $l(e_1) \ldots l(e_i)$ can be joined together with a suffix $l(e^i_l) \ldots l(e^i_n)$ such that for the whole word holds that $\delta^*(q_0, l(e_1) \ldots l(e_i) l(e^i_l) \ldots l(e^i_n)) \in F$.
4. At this point, we have to be careful which policy-compliant flows crossing the backward section $[e_j, \ldots, e_k]$ we are going to cancel. Indeed, as the first section $[e_1, \ldots, e_i]$ is glued together with

some policy-compliant path $p_i \in P$ at vertex u_j via condition (20), we are looking for states $q_{v_k} \in Q$ at vertex v_k such that, instead of (17) the following formula holds:

$$\exists q_{v_k} \in Q : \delta^*(q_{v_k}, l((v_k, u_k)) \dots l((v_j, u_j))) = q^i_{u_j}. \tag{21}$$

Indeed, after crossing section $[e_j, \dots, e_k]$ in reverse direction we want to end in state $q^i_{u_j}$ instead of q_{u_j}.

5. Having fulfilled the condition to allow canceling flow across the backward section of the augmenting path, we now need to check whether the flow arriving in v_k in state q_{v_k} can actually be diverted via the last forward section $[e_l, \dots, e_n]$ to the sink t reaching state q_t which needs to be a final state. This is accurately expressed by (18) which we can leave unmodified.
6. Suppose conditions (18), (20) and (21) have been fulfilled. Then calculate the residual capacity (19). Update flow f via (9) across the augmenting path using the value for the residual capacity, and update transitions-function T across the augmenting path as done previously.
7. Finally, we need to update the transitions-function T across the final part of the policy-compliant path p_i, cf. (20). This final part consists of the edges $[e^i_l, \dots, e^i_n]$, and thus we need to update T for all edges (u_x, u_y) in this partial path. We update T via the formulae $T'(u_x, u_y, q^i_{u_x}, q^i_{u_y}) = T(u_x, u_y, q^i_{u_x}, q^i_{u_y}) - c_f(p)$ and $T'(u_x, u_y, q_{u_x}, q_{u_y}) = T(u_x, u_y, q_{u_x}, q_{u_y}) + c_f(p)$, where we use the notation $q^i_{u_x}$ for the state that is reached at vertex u_x when starting in state $q^i_{u_j}$ in vertex u_j, while q_{u_x} denotes the state that is reached at vertex u_x when starting in state q_{u_j} in vertex u_j. After these updates, the augmenting path is processed entirely, and a new iteration can start.

Note that, in comparison to the approach for obtaining the first lower bound, condition (20) not only has merits for being more relaxed, it also carries a serious drawback as checking whether condition (20) holds involves more computational work than (16). One possible approach to check whether (20) actually holds, is by a depth-first walkthrough of the realization associated with the flow f. However, this is complicated by the fact that the flow might have multiple valid realizations, some of which are compatible with (20), while some are not. We discuss this now in detail.

1. Consider the augmenting path $p = [[e_1, \dots, e_i], [(u_j, v_j), \dots, (u_k, v_k)], [e_l, \dots, e_n]]$ from above, where the state at u_j is defined as $q_{u_j} = \delta^*(q_0, l(e_1) \dots l(e_i))$. Suppose there exist two policy-compliant paths $p_i \in P$ and $p_j \in P$, such that their states in vertex u_j are $q^i_{u_j}$ and $q^j_{u_j}$ respectively. Suppose that the realization of the flow that gave rise to these paths p_i and p_j is such that $p_i = [\dots, (u_j, u_1), (u_1, u_3), (u_3, u_4), (u_4, t)]$ and $p_j = [\dots, (u_j, u_2), (u_2, u_3), (u_3, u_5), (u_5, t)]$ and moreover that $l((u_j, u_1)) = a, l((u_1, u_3)) = b, l((u_3, u_4)) = c, l((u_4, t)) = d, l((u_j, u_2)) = e$, $l((u_2, u_3)) = f, l((u_3, u_5)) = g$ and finally $l((u_5, t)) = h$ with $\{a, b, c, d, e, f, g, h\} \subseteq \Sigma$. Then, as p_i and p_j are policy-compliant paths in the set P, it directly follows that $\delta^*(q^i_{u_j}, abcd) \in F$ and $\delta^*(q^j_{u_j}, efgh) \in F$, see Figure 3.

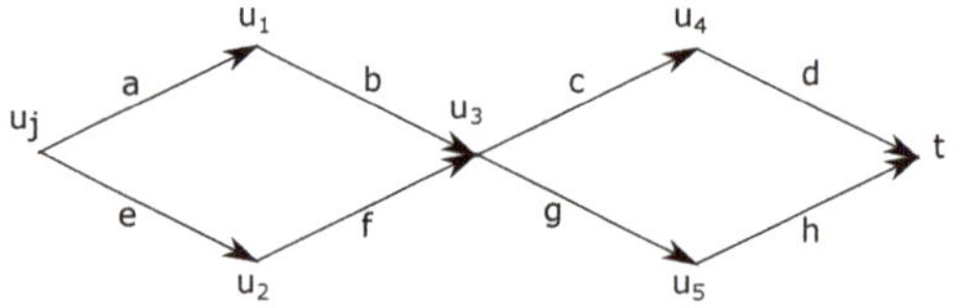

Figure 3. Fulfilling condition (20).

2. Suppose that moreover it holds that $\delta^*(q^i_{u_j}, abgh) \in F$ and $\delta^*(q^j_{u_j}, efcd) \in F$, that is, for the paths p_i and p_j it doesn't really matter which way they take in the second diamond. However, suppose that $\delta^*(q_{u_j}, abgh) \in F$ as well as $\delta^*(q_{u_j}, efcd) \in F$ but $\delta^*(q_{u_j}, abcd) \notin F$ and also $\delta^*(q_{u_j}, efgh) \notin F$.

3. Now, there is no way to fulfill condition (20), as no policy-compliant path through u_j that is part of the realization of f allows the prefix $l(e_1)\dots l(e_i)$ to be extended by a suffix to arrive at a final state in the sink. However, if the realization of the flow f was such that it contained the policy-compliant paths $[\dots,(u_j,u_1),(u_1,u_3),(u_3,u_5),(u_5,t)]$ and $[\dots,(u_j,u_2),(u_2,u_3),(u_3,u_4),(u_4,t)]$, then the prefix of the augmenting path p could have been extended via a matching suffix to a final state. In short, condition (20) would have been fulfilled.

Thus, although the relaxed condition (20) can be used to generate augmenting paths that would go unnoticed if applying (16), it suffers from the fact that previous decisions (like how to route policy-compliant paths in case of multiple possibilities) can have a large impact on the number of augmenting paths that can be found later on. As such, we have no guarantee that this second approach will always calculate the exact value for the policy-compliant maximum flow problem, but merely a lower bound. However, as (20) and (21) are fulfilled everytime (16) and (17) are fulfilled, we can immediately derive that the second lower bound will be tighter than the first lower bound.

4.6. Third Lower Bound

Considering the approaches taken to calculate the first and the second lower bound, observe that at several times choices have to be made, e.g., when fulfilling condition (17). Apart from the fact that some of the conditions might be overly restrictive, we face the fact that we might miss augmenting paths during the search when we make a bad choice, as well as the fact that we might need additional iterations to reach the point where no more augmenting paths can be found. For example, consider (17) and suppose that multiple states exist that fulfill the condition. By choosing one of them, an already present policy-compliant path is chosen too, which implies that the amount of flow that can be canceled has an upper bound that is equal to the minimum value of residual capacity across the backward section. In turn, this restricts the total amount of flow that can be sent over the augmenting path due to (19).

Thus, in order to allow as much flow as possible to be sent over the augmenting path that is constructed, one can check at the same time whether other transitions can be applied in parallel, such that the total amount of flow that is canceled across the backward section is maximized. First, we want to allow multiple pre-existing policy-compliant paths that cross the cut-vertex to be involved in the process, such that we do not have to choose at all which path we want to work with in (20). Multiple policy-compliant paths might fulfill the condition, even reaching the same state in the cut-vertex but following another way to the sink. Also, we want to allow multiple states at the end of the backward section to be involved in the process, such that we do not have to choose at all which state we want to work with in (21). Multiple states might fulfill the condition, and they even might be equal to each other even though they are part of different policy-compliant paths. We need to match the possibilities in both sets to maximize the flow that we can cancel along the backward section in the middle of the augmenting path. This being, of course, under the additional condition (18).

We start again with augmenting path p, set P and set S. Now select a subset $P' \subseteq P$ such that

$$\forall p_i \in P' : \delta^*(q_{u_j}, l(e^i_l)\dots l(e^i_n)) \in F \tag{22}$$

$$\forall p_i \in P' : \delta^*(q^i_{v_k}, l((v_k,u_k))\dots l((v_j,u_j))) = q^i_{u_j} \tag{23}$$

$$\forall p_i \in P' : \delta^*(q^i_{v_k}, l(e_l)\dots l(e_n)) \in F \tag{24}$$

$$\Sigma_{p_i \in P'} c_i \leq min(c_{u_j}, c_{v_k}). \tag{25}$$

This set P' can be iteratively built, starting from the empty set and choosing and adding p_i's while (25) holds. Once the iteration comes to a stop, augment the flow along the path p with the value $\Sigma_{p_i \in P'} c_i$ and update f and T. As the conditions (18), (20) and (21) imply that (22)–(24) will also hold, we obtain once more a tighter bound. However, there is no guarantee that the policy-compliant maximum flow will be reached. This is the most generic specification of a solution to the policy-compliant maximum flow problem that we derived.

4.7. Implementation and Experimental Results

Implementing classical maximum flow algorithms is not difficult as they are well-studied. Typical approaches include augmenting path algorithms, push-relabel algorithms and others. They have varying complexities in solving the problem. In practice, execution times are also dependent on the structure and density of the network. We chose an augmenting path approach. Finite state automata are typically implemented in a table-driven way, and it is quite straightforward to implement them efficiently, like we did. However, combining a maxflow algorithm with an FSA multiplies the complexities of both. Indeed, sending a flow over an edge multiple times was now allowed, as the states at the beginning and the end of the edge when crossing it for the first time might be different from the states while crossing it a second time. Thus, utmost care is needed to implement these algorithms correctly. All code is freely available in Supplementary Materials.

Although this paper is mainly intended as a theoretical study, we performed a large number of experiments validating our approach and formulas, in order to build some insights in the practical use of our approach, execution times and the accuracy reached. However, these values must be interpreted as rough estimates, and can only serve as rules-of-thumb, as much is dependent on the actual networks and finite state automata used in the experiments. Indeed, for evaluating the work, a network generator was implemented which allowed for the use of random networks having differing numbers of vertices and edges. Of course, the higher the density of edges, the more paths can be found and the higher the calculation time for the varying parts of the algorithms. We experimented with graphs having up to 256 vertices, with a random choice of edges (from one to the maximum number, i.e., $n(n-1)/2$). Also, we implemented an FSA generator, which allowed experimenting with different policies, all expressed as regular expressions. The maximal number of states allowed was maximally of the same order as the number of vertices, and clearly it holds that allowing more states leads to higher computational times. Indeed, a flow can reach a vertex in about $O(|V|)$ possible states, which also means that an algorithm looking for augmenting paths can be in about $O(|V|^2)$ states itself. Also, the flow over any edge is now parameterized by the states of both endpoints of the edge; as we use up to $|V|$ possible states this results in the memory-use for function T scaling in the order $O(|V|^4)$.

For our experiments with networks with up to 32 vertices and FSAs with up to 32 states, we found that running the brute-force algorithm either finished within a second, or did not finish even after 30 min, after which we always stopped the experiments. Careful investigation shows that the density of the network, combined with the structure of the FSA, is of prime importance to whether or not the algorithm finishes fast. For all the randomly generated experiments where the brute-force approach finished and provided us with the maximal value, the first lower bound also reached that same maximal value, but even in the worst observed cases in less than a millisecond, which is three orders of magnitude faster than the brute-force approach. The first lower bound algorithm could not find the maximal value only for specifically designed networks and FSAs, such as the network in Figure 3, aimed specifically at fooling the algorithm. For random experiments, the second lower bound algorithm always gives the same result as the first algorithm, taking only a little bit more time to execute. The third lower bound tries out much more combinations and comes closer to the brute-force approach in computational time, and in our random experiments always obtained the same maximal flow values as the second lower bound. We also executed random experiments with up to 256 vertices without having the maximal flow value, because the brute-force algorithm did not terminate. For the subset of these experiments where the three lower bound algorithms did terminate, also no differences were obtained in the resulting values. Thus, differences in the accuracy of results were never detected in any random experiment, and could only be obtained with specifically designed and contrived networks and automata. Runtimes for the first lower bound algorithm, for random networks having up to 256 vertices, are shown in Figure 4. If the algorithm finished, it typically finished fast (i.e., within about 10 s), as can be seen in the figure. If it didn't finish, we aborted the execution after 30 min (not depicted).

Concluding, we advise to use the first lower bound algorithm, as it terminates very fast and for our random experiments always reached the maximal value when known, i.e., when the brute-force approach also finished. It couldn't find the maximal value only in contrived examples, specifically designed to fool the algorithm. Also, we did not encounter random experiments where the three algorithms gave different results, so there is no reasonable situation where one might recommend using the second or third algorithm. We would like to stress again that the execution times we obtained varied wildly, being highly dependent on the structure of the network, the accompanying FSA and the combinatorial combination of the two.

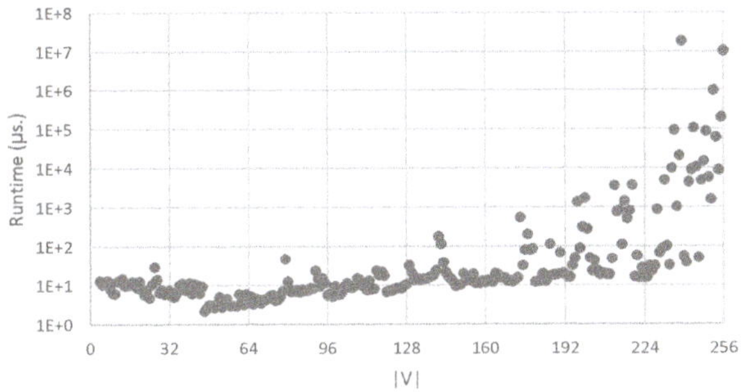

Figure 4. Runtimes for the first lower bound algorithm, for random networks with up to 256 vertices.

5. Conclusions and Future Work

Knowing the maximum throughput between two nodes in a network is key knowledge for a network operator. In practice, network policies severely complicate this task and often leave the question open-ended. We put forward a formal model with which it is possible to model real-life policy constraints, and we analyzed the impact of policies on throughput. As exact solutions are difficult to obtain, we defined a series of conditions and algorithms that allow us to calculate a sequence of increasingly tighter bounds on the exact value. These algorithms are built upon classic algorithms like Ford–Fulkerson to solve the generic maximum flow problem, which are adapted to our needs and augmented with specific functions to ease bookkeeping of additional data like transitions or the specific realization associated with the flow.

Although the approach to specify the conditions on the augmenting paths in order to enforce their policy-compliance is reasonable in combination with Ford-Fulkerson, it requires complex formulae that express conditions on the augmenting path, end-to-end. Future work in the direction of push-relabel algorithms would solve this problem: on the basis of local conditions, it might be possible to express specific constraints that guarantee path-compliancy upon arrival of the flow.

Supplementary Materials: All code is freely available and downloadable at http://www.dna.idlab.ugent.be.

Author Contributions: Conceptualization and methodology: P.A., D.C. and M.P.; software, validation, formal analysis, resources, visualization, writing: P.A.; investigation, writing—review and editing: P.A., D.C. and M.P.

Funding: This research was funded by Ghent University—imec, IOP 2016 Derudder-Pickavet.

Acknowledgments: The authors wish to thank Ghent University—imec for the support.

Conflicts of Interest: The authors declare no conflict of interest.

Abbreviations

The following abbreviations are used in this manuscript:

FSA	Finite state automaton
PCMF	Policy-compliant maximum flow

References

1. Gao, L. On Inferring Autonomous System Relationships in the Internet. *IEEE/ACM Trans. Netw.* **2001**, *9*, 733–745. [CrossRef]
2. Gao, L.; Rexford, J. Stable Internet Routing Without Global Coordination. *IEEE/ACM Trans. Netw.* **2001**, *9*, 681–692. [CrossRef]
3. Ford, D.R.; Fulkerson, D.R. *Flows in Networks*; Princeton University Press: Princeton, NJ, USA, 2010.
4. Cormen, T.H.; Leiserson, C.E.; Rivest, R.L.; Stein, C. *Introduction to Algorithms*, 3rd ed.; The MIT Press: Cambridge, MA, USA, 2009.
5. Hopcroft, J.E.; Ullman, J.D. *Formal Languages and Their Relation to Automata*; Addison-Wesley Longman Publishing Co., Inc.: Boston, MA, USA, 1969.
6. Linz, P. *An Introduction to Formal Language and Automata*; Jones and Bartlett Publishers, Inc.: Burlington, MA, USA, 2006.
7. Caesar, M.; Rexford, J. BGP Routing Policies in ISP Networks. *IEEE Netw.* **2005**, *19*, 5–11. doi:10.1109/MNET.2005.1541715. [CrossRef]
8. Feamster, N.; Borkenhagen, J.; Rexford, J. Guidelines for Interdomain Traffic Engineering. *SIGCOMM Comput. Commun. Rev.* **2003**, *33*, 19–30. [CrossRef]
9. Hu, C.; Chen, K.; Chen, Y.; Liu, B. Evaluating Potential Routing Diversity for Internet Failure Recovery. In Proceedings of the 29th Conference on Information Communications, San Diego, CA, USA, 14–19 March 2010; IEEE Press: Piscataway, NJ, USA, 2010; pp. 321–325.
10. Klöti, R.; Kotronis, V.; Ager, B.; Dimitropoulos, X. Policy-Compliant Path Diversity and Bisection Bandwidth. In Proceedings of the 2015 IEEE Conference on Computer Communications (INFOCOM), Kowloon, Hong Kong, China, 26 April–1 May 2015.
11. Sobrinho, J.L.; Quelhas, T. A Theory for the Connectivity Discovered by Routing Protocols. *IEEE/ACM Trans. Netw.* **2012**, *20*, 677–689. [CrossRef]
12. Erlebach, T.; Hall, A.; Panconesi, A.; Vukadinovic, D. Cuts and Disjoint Paths in the Valley-Free Model. *Internet Math.* **2007**, *3*, 333–359. [CrossRef]
13. Soulé, R.; Basu, S.; Kleinberg, R.; Sirer, E.G.; Foster, N. Managing the Network with Merlin. In Proceedings of the Twelfth ACM Workshop on Hot Topics in Networks, College Park, MD, USA, 21–22 November 2013; ACM: New York, NY, USA, 2013; pp. 24:1–24:7. [CrossRef]
14. Hinrichs, T.L.; Gude, N.S.; Casado, M.; Mitchell, J.C.; Shenker, S. Practical Declarative Network Management. In Proceedings of the 1st ACM Workshop on Research on Enterprise Networking (WREN '09), Barcelona, Spain, 21 August 2009; ACM: New York, NY, USA, 2009; pp. 1–10. [CrossRef]
15. Raghavan, B.; Verkaik, P.; Snoeren, A.C. Secure and Policy-compliant Source Routing. *IEEE/ACM Trans. Netw.* **2009**, *17*, 764–777. [CrossRef]
16. Godfrey, P.B.; Ganichev, I.; Shenker, S.; Stoica, I. Pathlet Routing. *SIGCOMM Comput. Commun. Rev.* **2009**, *39*, 111–122. [CrossRef]
17. Batista, B.; de Campos, G.; Fernandez, M. A proposal of policy based OpenFlow network management. In Proceedings of the 2013 20th International Conference on Telecommunications (ICT), Casablanca, Morocco, 6–8 May 2013; pp. 1–5. [CrossRef]

18. Lopes Alcantara Batista, B.; Lima de Campos, G.; Fernandez, M. Flow-based conflict detection in OpenFlow networks using first-order logic. In Proceedings of the 2014 IEEE Symposium on Computers and Communication (ISCC), Funchal, Portugal, 23–26 June 2014; pp. 1–6. [CrossRef]
19. Xu, W.; Rexford, J. MIRO: Multi-path Interdomain Routing. *SIGCOMM Comput. Commun. Rev.* **2006**, *36*, 171–182. [CrossRef]

applied sciences

MDPI

Article

Transient-Flow Modeling of Vertical Fractured Wells with Multiple Hydraulic Fractures in Stress-Sensitive Gas Reservoirs

Ping Guo [1,*], Zhen Sun [1], Chao Peng [2], Hongfei Chen [3] and Junjie Ren [1,4]

1 State Key Laboratory of Oil and Gas Reservoir Geology and Exploitation, Southwest Petroleum University, Chengdu 610500, China; zhensunswpu@hotmail.com (Z.S.); renjunjie1900@126.com (J.R.)
2 PetroChina Coalbed Methane Company Limited, Beijing 100083, China; chaopengpetrochina@outlook.com
3 Research Institute of Exploration & Development of Changqing Oilfield Company, PetroChina, Xi'an 710018, China; chenhongfeipc@hotmail.com
4 School of Sciences, Southwest Petroleum University, Chengdu 610500, China
* Correspondence: guopingswpi@vip.sina.com; Tel.: +86-028-83032346

Received: 18 February 2019; Accepted: 28 March 2019; Published: 31 March 2019

Abstract: Massive hydraulic fracturing of vertical wells has been extensively employed in the development of low-permeability gas reservoirs. The existence of multiple hydraulic fractures along a vertical well makes the pressure profile around the vertical well complex. This paper studies the pressure dependence of permeability to develop a seepage model of vertical fractured wells with multiple hydraulic fractures. Both transformed pseudo-pressure and perturbation techniques have been employed to linearize the proposed model. The superposition principle and a hybrid analytical-numerical method were used to obtain the bottom-hole pseudo-pressure solution. Type curves for pseudo-pressure are presented and identified. The effects of the relevant parameters (such as dimensionless permeability modulus, fracture conductivity coefficient, hydraulic-fracture length, angle between the two adjacent hydraulic fractures, the difference of the hydraulic-fracture lengths, and hydraulic-fracture number) on the type curve and the error caused by neglecting the stress sensitivity are discussed in detail. The proposed work can enrich the understanding of the influence of the stress sensitivity on the performance of a vertical fractured well with multiple hydraulic fractures and can be used to more accurately interpret and forecast the transient pressure.

Keywords: gas flow; stress-sensitive porous media; multiple hydraulic fractures; vertical fractured well

1. Introduction

Gas flow in porous media has recently attracted much attention and stimulated great interest in the development of oil and gas reservoirs [1–15]. In the past decades, many gas reservoirs with stress-sensitive permeability have been discovered and developed around the world. During the development process of stress-sensitive reservoirs, the continuously decreasing formation pressure leads to the decrease of the formation permeability. Therefore, it is critical to investigate the effect of the stress sensitivity of the permeability on the production. Recently, lots of models have been established to study the performance of various wells in stress-sensitive reservoirs [16–18].

On the other hand, in order to obtain economic benefit, hydraulic fracturing has been widely applied in the development of low-permeability gas reservoirs. A great variety of seepage models of vertical fractured wells have been proposed and the characteristic of pressure response has been studied in detail. Gringarten et al. [19] established a seepage model of a vertical fractured well with two symmetrical infinite-conductivity hydraulic fractures, which can be used to identify the linear flow. Later, by considering the effect of the fluid flow within the hydraulic fractures, a seepage model of a vertical fractured well with two symmetrical finite-conductivity hydraulic fractures was established [20,21], which can identify the bilinear flow. Some researchers pointed out the existence of two asymmetrical hydraulic fractures in practice and established some seepage models of vertical fractured wells with two asymmetrical hydraulic fractures [22–27]. In fact, massive hydraulic fracturing usually creates fractured network or several main hydraulic fractures near the wellbore instead of two symmetrical/asymmetrical hydraulic fractures [28–31]. Recently, seepage models of vertical fractured wells with multiple hydraulic fractures have received more attention. Shaoul et al. [32] proposed a numerical model of a vertical fractured well with multiple hydraulic fractures and investigated the pressure behavior of a vertical fractured well. Restrepo and Tiab [33] and Wang et al. [34] established semi-analytical models of a vertical fractured well with multiple infinite-conductivity hydraulic fractures. Ren and Guo [35] and Luo and Tang [36] proposed semi-analytical models of a vertical fractured well with multiple finite-conductivity hydraulic fractures. Compared with the pressure behavior of vertical fractured wells with two hydraulic fractures, the pressure response of vertical fractured wells with multiple hydraulic fractures is very complex. Although the effect of relevant parameters on the pressure response of vertical fractured wells with multiple hydraulic fractures has been studied, and the characteristics of the type of curves is not clearly understood. More important, little work has focused on vertical fractured wells with multiple hydraulic fractures in stress-sensitive gas reservoirs. In particular, the effect of relevant parameters on the error caused by neglecting the stress sensitivity has not been recognized clearly.

In this paper, we propose a seepage model of vertical fractured wells with multiple finite-conductivity hydraulic fractures. Type curves are identified and analyzed. The effects of relevant parameters on type curves and the error caused by neglecting the stress sensitivity are discussed in detail. This work can enrich the understanding of the influence of stress sensitivity on the performance of vertical fractured wells with multiple hydraulic fractures.

2. Physical Model

Complex geometry patterns of hydraulic fractures for vertical fractured wells can be described by the multiple-fracture model which is shown in Figure 1. Some assumptions of the present model are given as follows:

(1) The gas reservoir is a homogeneous and laterally infinite formation with a constant thickness. The top and bottom boundaries of the gas reservoir are assumed to be impermeable.

(2) The finite-conductivity hydraulic fractures emanate from the wellbore in the horizontal plane (as shown in Figure 1b) and completely penetrate the formation in the vertical plane. The length of each hydraulic fracture may be different.

(3) The vertical fractured well produces only from the hydraulic fractures and the production rate of the vertical fractured well keeps constant.

(4) The permeabilities of both the reservoir and hydraulic fractures change with pressure.

(5) The gas reservoir has a constant temperature and uniform initial pressure.

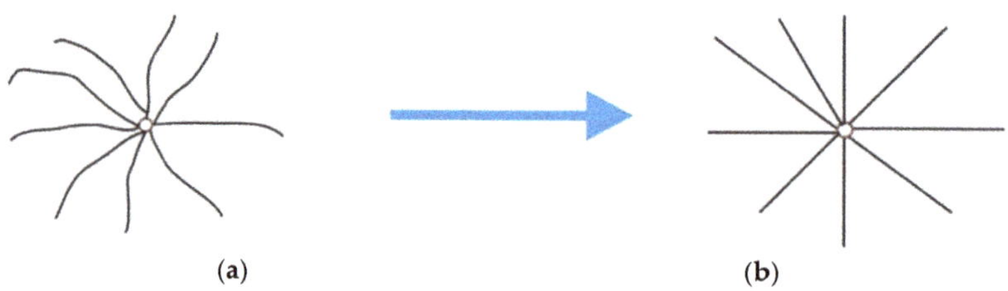

Figure 1. Vertical fractured well with multiple hydraulic fractures and its model. (**a**) Complex geometry pattern, (**b**) Multiple fracture model.

3. Mathematical Model

A semi-analytical model of vertical fractured well with multiple finite-conductivity hydraulic fractures in stress-sensitive gas reservoirs is established and solved by incorporating various methods and techniques. The nomenclatures of the present model are listed in Appendix A. The flow chart of solution methodology is shown in Figure 2.

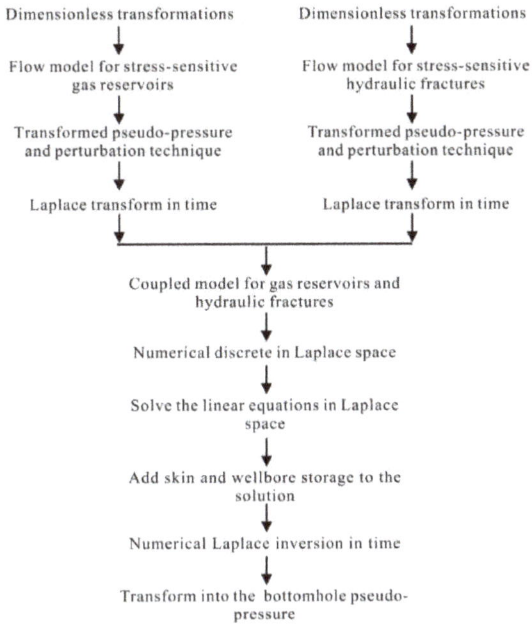

Figure 2. Flow chart of the solution methodology.

3.1. Flow Model for Stress-Sensitive Gas Reservoirs

The reservoir permeability changes with reservoir pressure and can be described as follows [37]:

$$k = k_{\mathrm{i}} e^{-\gamma(\psi_{\mathrm{i}} - \psi)}, \tag{1}$$

where ψ is the pseudo-pressure, which is defined as

$$\psi = \int_0^p \frac{2p}{\mu Z} \mathrm{d}p. \tag{2}$$

The governing equation of gas flow in stress-sensitive porous media is given as follows [37]:

$$\frac{\partial^2 \psi}{\partial r^2} + \frac{1}{r}\frac{\partial \psi}{\partial r} + \gamma \left(\frac{\partial \psi}{\partial r}\right)^2 = e^{\gamma(\psi_{\mathrm{i}} - \psi)} \frac{\phi \mu c_{\mathrm{t}}}{k_{\mathrm{i}}} \frac{\partial \psi}{\partial t}. \tag{3}$$

Both gas viscosity μ and total compressibility c_{t} in Equation (3) are functions of pressure. In order to simplify the proposed model and obtain an analytical solution, many research papers [26,37,38] treat the values of μ and c_{t} in Equation (3) as constants, which could make the simulation results be accurate enough for engineering requirements. Following the suggestion by Wang [37], the values of μ and c_{t} are evaluated at the initial condition, i.e., $\mu = \mu(p_{\mathrm{i}})$ and $c_{\mathrm{t}} = c_{\mathrm{t}}(p_{\mathrm{i}})$ in this paper.

With the assumption of the uniform initial pressure, the initial condition is expressed as

$$\psi(r, t=0) = \psi_{\mathrm{i}}. \tag{4}$$

The gas reservoir is assumed to be laterally infinite, and thus, the outer boundary condition is

$$\psi(r \to \infty, t) = \psi_{\mathrm{i}}. \tag{5}$$

The inner boundary condition for a point source can be written as

$$\lim_{r \to 0} e^{-\gamma(\psi_{\mathrm{i}}-\psi)} r \frac{\partial \psi}{\partial r} = \frac{p_{\mathrm{sc}} q(r=0,t) T}{\pi k_{\mathrm{i}} h T_{\mathrm{sc}}}. \tag{6}$$

Introducing the dimensionless variables listed in Table 1, Equations (3)–(6) are rewritten as

$$\frac{\partial^2 \psi_{\mathrm{D}}}{\partial r_{\mathrm{D}}^2} + \frac{1}{r_{\mathrm{D}}} \frac{\partial \psi_{\mathrm{D}}}{\partial r_{\mathrm{D}}} - \gamma_{\mathrm{D}} \left(\frac{\partial \psi_{\mathrm{D}}}{\partial r_{\mathrm{D}}} \right)^2 = e^{\gamma_{\mathrm{D}} \psi_{\mathrm{D}}} \frac{\partial \psi_{\mathrm{D}}}{\partial t_{\mathrm{D}}}, \tag{7}$$

$$\psi_{\mathrm{D}}(r_{\mathrm{D}}, t_{\mathrm{D}}=0) = 0, \tag{8}$$

$$\psi_{\mathrm{D}}(r_{\mathrm{D}}, t_{\mathrm{D}}=0) = 0, \tag{9}$$

$$\lim_{r_{\mathrm{D}} \to 0} e^{-\gamma_{\mathrm{D}} \psi_{\mathrm{D}}} r_{\mathrm{D}} \frac{\partial \psi_{\mathrm{D}}}{\partial r_{\mathrm{D}}} = -q_{\mathrm{D}}(r_{\mathrm{D}}=0, t_{\mathrm{D}}). \tag{10}$$

Table 1. Definitions of dimensionless variables.

Nomenclature	Definition
Dimensionless pseudo − pressure	$\psi_{\mathrm{D}} = \frac{\pi k_{\mathrm{i}} h T_{\mathrm{sc}}}{Q_{\mathrm{sc}} p_{\mathrm{sc}} T}(\psi_{\mathrm{i}} - \psi)$
Dimensionless transformed pseudo − pressure	$\xi_{\mathrm{D}} = \frac{\pi k_{\mathrm{i}} h T_{\mathrm{sc}}}{Q_{\mathrm{sc}} p_{\mathrm{sc}} T}\xi$
Dimensionless time	$t_{\mathrm{D}} = \frac{k_{\mathrm{i}} t}{\phi c_{\mathrm{t}} \mu r_{\mathrm{w}}^2}$
Dimensionless distance	$r_{\mathrm{D}} = \frac{r}{r_{\mathrm{w}}}$
Dimensionless wellbore storage coefficient	$C_{\mathrm{D}} = \frac{C}{2\pi \phi c_{\mathrm{t}} h r_{\mathrm{w}}^2}$
Dimensionless flow rate	$q_{\mathrm{D}} = \frac{q}{Q_{\mathrm{sc}}}$
Dimensionless flow − rate density	$q_{\mathrm{fD}} = \frac{r_{\mathrm{w}} q_{\mathrm{f}}}{Q_{\mathrm{sc}}}$
Dimensionless permeability modulus	$\gamma_{\mathrm{D}} = \gamma \frac{Q_{\mathrm{sc}} p_{\mathrm{sc}} T}{\pi k_{\mathrm{i}} h T_{\mathrm{sc}}}$
Fracture conductivity coefficient	$C_{\mathrm{fD}} = \frac{k_{\mathrm{fi}} w_{\mathrm{f}}}{k_{\mathrm{i}} r_{\mathrm{w}}}$

The point source model including Equations (7)–(10) shows strong non-linearity which can be alleviated by introducing the following expression [16]:

$$\psi_{\mathrm{D}}(r_{\mathrm{D}}, t_{\mathrm{D}}) = -\frac{1}{\gamma_{\mathrm{D}}} \ln[1 - \gamma_{\mathrm{D}} \xi_{\mathrm{D}}(r_{\mathrm{D}}, t_{\mathrm{D}})]. \tag{11}$$

The perturbation technique, Laplace transform, and superposition principle [39] were employed to obtain the pressure distribution in the gas reservoir with a vertical fractured well with multiple hydraulic fractures producing at a constant rate (see Appendix B):

$$\bar{\xi}_{\mathrm{D0}}(r_{\mathrm{D}}, \theta, s) = \sum_{j=1}^{n} \left[\int_0^{L_{fj\mathrm{D}}} \bar{q}_{f\mathrm{D}}(\alpha, \theta_{\mathrm{f}j}, s) \mathrm{K}_0\left(\sqrt{s}\sqrt{r_{\mathrm{D}}^2 + \alpha^2 - 2 r_{\mathrm{D}} \alpha \cos(\theta - \theta_{\mathrm{f}j})}\right) \mathrm{d}\alpha \right]. \tag{12}$$

3.2. Flow Model for Stress-Sensitive Hydraulic Fractures

The dimensionless model for gas flow within stress-sensitive hydraulic fractures is given as (see Appendix C)

$$\frac{\partial^2 \xi_{\mathrm{fD}}(x_{i\mathrm{D}}, t_{\mathrm{D}})}{\partial x_{i\mathrm{D}}^2} - \frac{2\pi}{C_{\mathrm{fD}}} q_{\mathrm{fD}}(x_{i\mathrm{D}}, t_{\mathrm{D}}) = 0, \ (0 < x_{i\mathrm{D}} < L_{\mathrm{fiD}}), \tag{13}$$

$$\left. \frac{\partial \xi_{\mathrm{fD}}(x_{i\mathrm{D}}, t_{\mathrm{D}})}{\partial x_{i\mathrm{D}}} \right|_{x_{i\mathrm{D}}=0} = -\frac{2\pi}{C_{\mathrm{fD}}} Q_{i\mathrm{D}}(t_{\mathrm{D}}). \tag{14}$$

$$Q_{i\mathrm{D}}(t_{\mathrm{D}}) = \int_0^{L_{\mathrm{fiD}}} q_{fD}(x_{i\mathrm{D}}, t_{\mathrm{D}}) \mathrm{d}x_{i\mathrm{D}}, \tag{15}$$

$$\sum_{j=1}^{n} Q_{j\mathrm{D}}(t_{\mathrm{D}}) = 1. \tag{16}$$

According to the relationship between Cartesian coordinate system (x_i, y_i) and polar coordinate system (r, θ) (see Figure 3), Equations (13)–(16) can be rewritten in the polar coordinate system as follows:

$$\frac{\partial^2 \xi_{\mathrm{fD}}(r_{\mathrm{D}}, \theta_{\mathrm{fi}}, t_{\mathrm{D}})}{\partial r_{\mathrm{D}}^2} - \frac{2\pi}{C_{\mathrm{fD}}} q_{\mathrm{fD}}(r_{\mathrm{D}}, \theta_{\mathrm{fi}}, t_{\mathrm{D}}) = 0, \ (0 < r_{\mathrm{D}} < L_{\mathrm{fiD}}), \tag{17}$$

$$\left. \frac{\partial \xi_{\mathrm{fD}}(r_{\mathrm{D}}, \theta_{\mathrm{fi}}, t_{\mathrm{D}})}{\partial r_{\mathrm{D}}} \right|_{r_{\mathrm{D}}=0} = -\frac{2\pi}{C_{\mathrm{fD}}} Q_{i\mathrm{D}}(t_{\mathrm{D}}), \tag{18}$$

$$Q_{i\mathrm{D}}(t_{\mathrm{D}}) = \int_0^{L_{\mathrm{fiD}}} q_{fD}(r_{\mathrm{D}}, \theta_{\mathrm{fi}}, t_{\mathrm{D}}) \mathrm{d}r_{\mathrm{D}}, \tag{19}$$

$$\sum_{j=1}^{n} Q_{j\mathrm{D}}(t_{\mathrm{D}}) = 1. \tag{20}$$

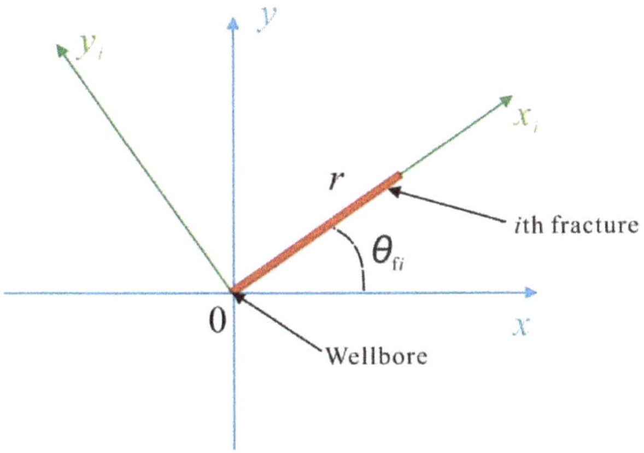

Figure 3. Schematic of different coordinate systems.

Integrating Equation (17) from 0 to x' with respect to r_{D} yields

$$\frac{\partial \xi_{\mathrm{fD}}(x', \theta_{\mathrm{fi}}, t_{\mathrm{D}})}{\partial r_{\mathrm{D}}} - \frac{\partial \xi_{\mathrm{fD}}(0, \theta_{\mathrm{fi}}, t_{\mathrm{D}})}{\partial r_{\mathrm{D}}} = \frac{2\pi}{C_{\mathrm{fD}}} \int_0^{x'} q_{fD}(x'', \theta_{\mathrm{fi}}, t_{\mathrm{D}}) \mathrm{d}x''. \tag{21}$$

Substituting Equations (18) and (19) into Equation (21) yields

$$\frac{\partial \xi_{\mathrm{fD}}(x', \theta_{\mathrm{fi}}, t_{\mathrm{D}})}{\partial r_{\mathrm{D}}} + \frac{2\pi}{C_{\mathrm{fD}}}\int_0^{L_{\mathrm{fiD}}} q_{fD}(r_{\mathrm{D}}, \theta_{\mathrm{fi}}, t_{\mathrm{D}})\mathrm{d}r_{\mathrm{D}} = \frac{2\pi}{C_{\mathrm{fD}}}\int_0^{x'} q_{fD}(x'', \theta_{\mathrm{fi}}, t_{\mathrm{D}})\mathrm{d}x''. \tag{22}$$

Integrating Equation (22) from 0 to r_{D} with respect to r_{D} yields

$$\xi_{\mathrm{wHD}}(t_{\mathrm{D}}) - \xi_{\mathrm{fD}}(r_{\mathrm{D}}, \theta_{\mathrm{fi}}, t_{\mathrm{D}}) = \frac{2\pi}{C_{\mathrm{fD}}}\left[r_{\mathrm{D}}\int_0^{L_{\mathrm{fiD}}} q_{fD}(r_{\mathrm{D}}, \theta_{\mathrm{fi}}, t_{\mathrm{D}})\mathrm{d}r_{\mathrm{D}} - \int_0^{r_{\mathrm{D}}}\int_0^{x'} q_{fD}(x'', \theta_{\mathrm{fi}}, t_{\mathrm{D}})\mathrm{d}x''\mathrm{d}x'\right], \tag{23}$$

where

$$\xi_{\mathrm{wHD}}(t_{\mathrm{D}}) = \xi_{\mathrm{fD}}(0, \theta_{\mathrm{fi}}, t_{\mathrm{D}}). \tag{24}$$

3.3. Coupled Discrete Model

The flow model for gas reservoirs can associate with the flow model for hydraulic fractures by the following relationship:

$$\psi_{\mathrm{fD}}(r_{\mathrm{D}}, \theta_{\mathrm{fi}}, t_{\mathrm{D}}) = \psi_{\mathrm{D}}(r_{\mathrm{D}}, \theta_{\mathrm{fi}}, t_{\mathrm{D}}),\ (0 < r_{\mathrm{D}} < L_{\mathrm{fiD}}). \tag{25}$$

Substituting Equations (11) and (A37) into Equation (25), one can derive

$$\xi_{\mathrm{fD}}(r_{\mathrm{D}}, \theta_{\mathrm{fi}}, t_{\mathrm{D}}) = \xi_{\mathrm{D}}(r_{\mathrm{D}}, \theta_{\mathrm{fi}}, t_{\mathrm{D}}). \tag{26}$$

The zero-order perturbation solution of the ξ_{D} is accurate enough in practice, and thus we can obtain

$$\xi_{\mathrm{D}}(r_{\mathrm{D}}, \theta_{\mathrm{fi}}, t_{\mathrm{D}}) = \xi_{\mathrm{D0}}(r_{\mathrm{D}}, \theta_{\mathrm{fi}}, t_{\mathrm{D}}). \tag{27}$$

Combining Equations (26) and (27) yields that

$$\xi_{\mathrm{fD}}(r_{\mathrm{D}}, \theta_{\mathrm{fi}}, t_{\mathrm{D}}) = \xi_{\mathrm{D0}}(r_{\mathrm{D}}, \theta_{\mathrm{fi}}, t_{\mathrm{D}}). \tag{28}$$

Substituting Equation (28) into Equation (23) leads to

$$\xi_{\mathrm{wHD}}(t_{\mathrm{D}}) - \xi_{\mathrm{D0}}(r_{\mathrm{D}}, \theta_{\mathrm{fi}}, t_{\mathrm{D}}) = \frac{2\pi}{C_{\mathrm{fD}}}\left[r_{\mathrm{D}}\int_0^{L_{\mathrm{fiD}}} q_{fD}(r_{\mathrm{D}}, \theta_{\mathrm{fi}}, t_{\mathrm{D}})\mathrm{d}r_{\mathrm{D}} - \int_0^{r_{\mathrm{D}}}\int_0^{x'} q_{fD}(x'', \theta_{\mathrm{fi}}, t_{\mathrm{D}})\mathrm{d}x''\mathrm{d}x'\right]. \tag{29}$$

Taking the Laplace transform of Equation (29), one can obtain

$$\overline{\xi}_{\mathrm{wHD}}(s) - \overline{\xi}_{\mathrm{D0}}(r_{\mathrm{D}}, \theta_{\mathrm{fi}}, s) = \frac{2\pi}{C_{\mathrm{fD}}}\left[r_{\mathrm{D}}\int_0^{L_{\mathrm{fiD}}} \overline{q}_{fD}(r_{\mathrm{D}}, \theta_{\mathrm{fi}}, s)\mathrm{d}r_{\mathrm{D}} - \int_0^{r_{\mathrm{D}}}\int_0^{x'} \overline{q}_{fD}(x'', \theta_{\mathrm{fi}}, s)\mathrm{d}x''\mathrm{d}x'\right]. \tag{30}$$

Substituting Equation (12) into Equation (30), one can derive

$$\begin{aligned} &\overline{\xi}_{\mathrm{wHD}}(s) - \sum_{j=1}^{n}\left[\int_0^{L_{fjD}} \overline{q}_{fD}(\alpha, \theta_{\mathrm{fj}}, s)\mathrm{K}_0\left(\sqrt{s}\sqrt{r_{\mathrm{D}}^2 + \alpha^2 - 2r_{\mathrm{D}}\alpha\cos(\theta_{\mathrm{fi}} - \theta_{\mathrm{fj}})}\right)\mathrm{d}\alpha\right] \\ &= \frac{2\pi}{C_{\mathrm{fD}}}\left[r_{\mathrm{D}}\int_0^{L_{\mathrm{fiD}}} \overline{q}_{fD}(r_{\mathrm{D}}, \theta_{\mathrm{fi}}, s)\mathrm{d}r_{\mathrm{D}} - \int_0^{r_{\mathrm{D}}}\int_0^{x'} \overline{q}_{fD}(x'', \theta_{\mathrm{fi}}, s)\mathrm{d}x''\mathrm{d}x'\right]. \end{aligned} \tag{31}$$

Employing the Laplace transform of Equations (19) and (20), one can obtain

$$\sum_{i=1}^{n}\int_0^{L_{\mathrm{fiD}}} \overline{q}_{fD}(r_{\mathrm{D}}, \theta_{\mathrm{fi}}, s)\mathrm{d}r_{\mathrm{D}} = \frac{1}{s}. \tag{32}$$

The coupled model consists of Equations (31) and (32), which can be solved by the numerical discrete method [40]. Each hydraulic fracture is discretized into some segments, and the flow rate in each segment is considered to stay the same at a certain time. The discrete schematic of a hydraulic fracture of a vertical fractured well is shown in Figure 4.

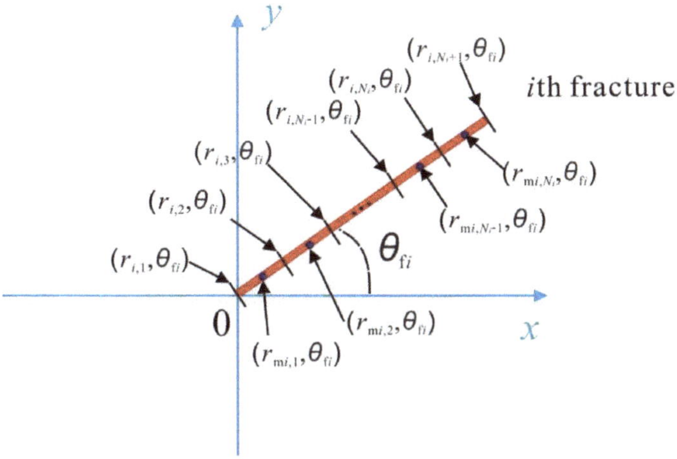

Figure 4. Discrete schematic of the *i*th hydraulic fracture of a vertical fractured well.

And then, the discrete forms of Equations (31) and (32) can be obtained as follows

$$\begin{aligned}&\bar{\xi}_{\mathrm{wHD}}(s)-\sum_{i=1}^{n}\sum_{j=1}^{N_i}\left[\bar{q}_{\mathrm{f}i,j\mathrm{D}}(s)\int_{r_{i,j\mathrm{D}}}^{r_{i,j+1\mathrm{D}}}\mathrm{K}_0\left(\sqrt{s}\sqrt{r_{\mathrm{m}k,v\mathrm{D}}^2+\alpha^2-2r_{\mathrm{m}k,v\mathrm{D}}\alpha\cos(\theta_{\mathrm{f}k}-\theta_{\mathrm{f}i})}\right)\mathrm{d}\alpha\right]+\\&\frac{2\pi}{C_{\mathrm{fD}}}\left\{\sum_{i=1}^{v-1}\bar{q}_{\mathrm{f}k,i\mathrm{D}}(s)\cdot\left[(v-i)\cdot\Delta L_{\mathrm{f}k\mathrm{D}}^2\right]+\frac{\Delta L_{\mathrm{f}k\mathrm{D}}^2}{8}\cdot\bar{q}_{\mathrm{f}k,v\mathrm{D}}(s)-r_{\mathrm{m}k,v\mathrm{D}}\sum_{i=1}^{i=N_k}\bar{q}_{\mathrm{f}k,i\mathrm{D}}(s)\Delta L_{\mathrm{f}k\mathrm{D}}\right\}=0,\end{aligned} \tag{33}$$

$$\sum_{i=1}^{n}\sum_{j=1}^{N_i}\left[\bar{q}_{\mathrm{f}i,j\mathrm{D}}(s)\Delta L_{\mathrm{f}i\mathrm{D}}\right]=\frac{1}{s}, \tag{34}$$

where $1\le k\le n, 1\le v\le N_k$ and $\Delta L_{\mathrm{f}i\mathrm{D}}=L_{\mathrm{f}i\mathrm{D}}/N_i$.

Equations (33) and (34) compose $\left(\sum_{i=1}^{n}N_i+1\right)$ linear equations with $\left(\sum_{i=1}^{n}N_i+1\right)$ unknowns. The unknowns are $\bar{\xi}_{\mathrm{wHD}}(s)$ and $\bar{q}_{\mathrm{f}k,v\mathrm{D}}(s)$ $(1\le k\le n, 1\le v\le N_k)$, which can be obtained by solving the linear equations, and then the wellbore storage and the skin near the wellbore can be taken into account by the following formula [41]

$$\bar{\xi}_{\mathrm{wD}}(s)=\frac{s\bar{\xi}_{\mathrm{wHD}}(s)+S_{\mathrm{f}}}{s+s^2C_{\mathrm{D}}\left[s\bar{\xi}_{\mathrm{wHD}}(s)+S_{\mathrm{f}}\right]}. \tag{35}$$

With the aid of the numerical inversion method [42], $\bar{\xi}_{\mathrm{wD}}(s)$ in Laplace space is transformed into $\xi_{\mathrm{wD}}(t_{\mathrm{D}})$ in real space

$$\xi_{\mathrm{wD}}(t_{\mathrm{D}})=\frac{\ln 2}{t_{\mathrm{D}}}\sum_{i=1}^{N}V_i\bar{\xi}_{wD}(s),\ (N=8), \tag{36}$$

where

$$s=\frac{i\ln 2}{t_{\mathrm{D}}}, \tag{37}$$

$$V_i = (-1)^{\frac{N}{2}+i} \sum_{k=\frac{i+1}{2}}^{\min(i,\frac{N}{2})} \frac{k^{\frac{N}{2}+1}(2k)!}{\left(\frac{N}{2}-k\right)!k!(k-1)!(i-k)!(2k-i)!}. \quad (38)$$

Finally, the dimensionless bottom-hole pseudo-pressure of a vertical fractured well with constant production rate is obtained by $\psi_{wD}(t_D) = -\ln[1 - \gamma_D \cdot \xi_{wD}(t_D)]/\gamma_D$.

4. Model Validation and Application

Because there is no analytical inversion solution of $\overline{\xi}_{wD}(s)$, it is difficult to discuss the accuracy of the Laplace transform by comparing the analytical inversion solution with the numerical inversion solution [43]. In order to validate the proposed model and show how this model was used in practice, the drawdown test data of a vertical fractured well with four fracture wings were collected from the published literature [44]. The drawdown test data in the literature [44] were generated by the reservoir simulator which is based on the implicit finite difference method. The vertical fractured well produces at a constant production rate of 0.1639 m^3/s. Basic parameters of the vertical fractured well are listed in Table 2. The proposed model is employed to simulate the bottom-hole pressure under the constant-rate-production condition. The simulated bottom-hole pseudo-pressure and bottom-hole pressure are compared with the results published in the literature [44]. It is shown from Figure 5 that the simulated bottom-hole pseudo-pressure and bottom-hole pressure by the proposed model are in good agreement with the bottom-hole pressure data published in the literature [44], which validates the proposed model.

Furthermore, the proposed model can be used to predict the bottom-hole pressure under the constant-rate-production condition. Based on the proposed model, it was easy to obtain the bottom-hole pressure of the vertical fractured well at any given time. For example, when the vertical fractured well produces at a constant production rate of 0.1639 m^3/s, the bottom-hole pressures are 41.132 MPa, 37.676 MPa, 31.484 MPa at $t = 100,\ 1000,\ 10000$ hours, respectively.

Table 2. Parameters of a vertical fractured well with four fracture wings in the published literature [44].

Parameter	Value
Reservoir thickness, h, m	7.62
Reservoir porosity, ϕ	0.075
Initial reservoir permeability, k_i, mD	0.05
Initial fracture permeability, k_{fi}, mD	3927
Initial reservoir pressure, p_i, MPa	44.785
Initial reservoir pseudo-pressure, ψ_i, $MPa^2/(mPa \cdot s)$	9.6321×10^4
Total compressibility, c_t, MPa^{-1}	1.43×10^{-2}
Wellbore radius, r_w, m	0.0762
Gas viscosity, μ, $mPa \cdot s$	0.0252
Reservoir temperature, T, K	412.04
Fracture width, w_f, m	0.0061
Well production rate, Q_{sc}, m^3/s	0.1639
Polar angles of the hydraulic fractures, θ_{fj}, degree	$0,\ \pi/2,\ \pi,\ 3\pi/2$
Lengths of the hydraulic fractures, L_{fj}, m	149.413, 2.987, 149.413, 2.987

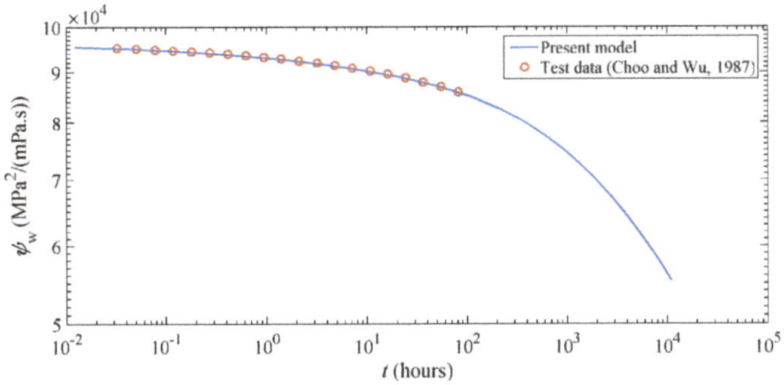

(a) bottom-hole pseudo-pressure

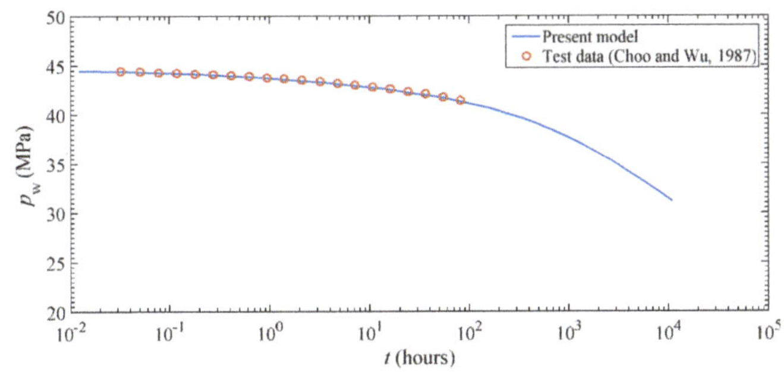

(b) bottom-hole pressure

Figure 5. Comparisons of bottom-hole pseudo-pressure and bottom-hole pressure between our simulation results and the test data [44]. (**a**) bottom-hole pseudo -pressure, (**b**) bottom-hole pressure

5. Results and Analysis

5.1. Type Curves and Flow Regimes

The bottom-hole pseudo-pressure of a vertical fractured well with multiple hydraulic fractures in stress-sensitive gas reservoirs was calculated by the proposed model. Type curves for the bottom-hole pseudo-pressure were plotted and the characteristics of the bottom-hole pseudo-pressure behavior was analyzed. The established model in this work was suitable for multiple hydraulic fractures with arbitrary fracture number, arbitrary fracture length, and arbitrary fracture orientation. In order to investigate the effect of the hydraulic-fracture distribution on the pseudo-pressure response, without loss of generality, we mainly focused on the vertical fractured well with four fracture wings which is shown in Figure 6. The first and second fracture wings were assumed to have the equal length (X_f), and the third and fourth fracture wings were set to be the equal length (Y_f).

Figure 7 shows the type curves for the bottom-hole pseudo-pressure of a vertical fractured well with multiple hydraulic fractures in stress-sensitive gas reservoirs. The type curves for the bottom-hole pseudo-pressure consist of the pseudo-pressure curve ($x-$axis: t_D/C_D, $y-$axis: ψ_{wD}) and the pseudo-pressure derivative curve ($x-$axis: t_D/C_D, $y-$axis: $\psi'_{wD} \cdot t_D/C_D$). It is observed that there are seven possible flow regions in the type curves as follows:

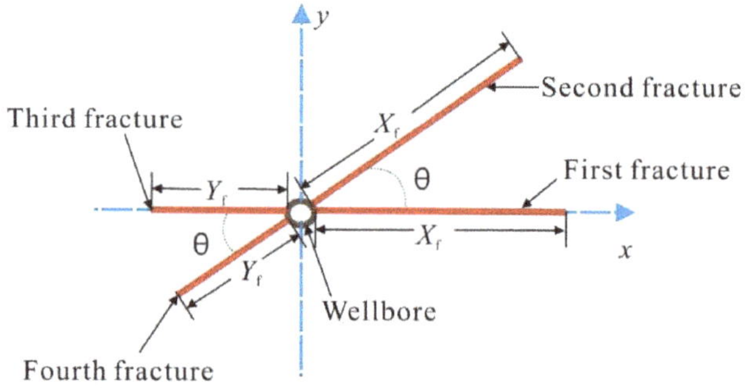

Figure 6. Schematic of a vertical fractured well with four fracture wings.

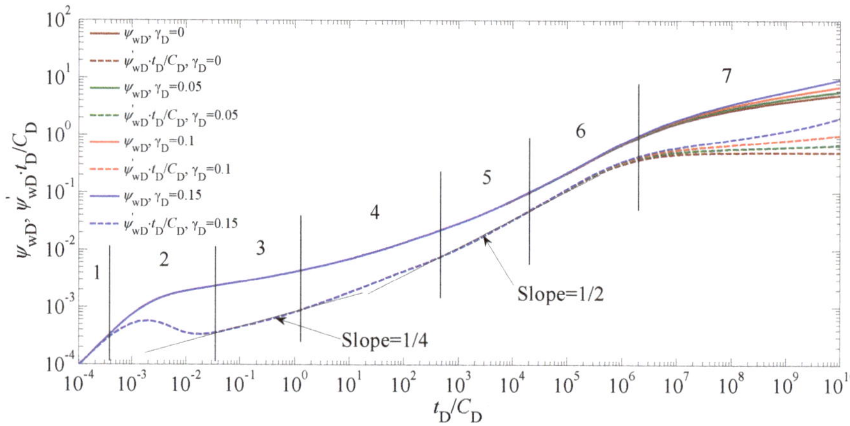

Figure 7. Type curves for the bottom-hole pseudo-pressure of a vertical fractured well with multiple hydraulic fractures in stress-sensitive gas reservoirs with different values of the γ_D ($n = 4$, $L_{f1} = L_{f2} = X_f = 20$ m, $L_{f3} = L_{f4} = Y_f = 80$ m, $\theta = 30°$, $C_{fD} = 5 \times 10^4$, $S_f = 10^{-3}$, $C_D = 0.1$).

(1) Wellbore storage period (WSP): The produced gas comes from the storage gas in the wellbore, and the gas in the reservoir has not flowed into the wellbore. Both the pseudo-pressure curve and the pseudo-pressure derivative curve exhibit as straight lines with the slope being one.

(2) Transitional flow after wellbore storage period (TFAWSP): the gas in the reservoir begins to flow into the wellbore. This flow period is controlled by both the wellbore storage and skin near the wellbore. The pseudo-pressure derivative curve appears as a "hump".

(3) Bilinear flow period (BFP): Linear flows take place within the hydraulic fractures and perpendicular to the hydraulic-facture surfaces in the reservoir simultaneously. The pseudo-pressure derivative curve exhibits as a straight line with the slope being 1/4.

(4) Transitional flow after bilinear flow period (TFABFP): The duration of this flow period is dependent on the difference of the hydraulic-fracture lengths. The shorter the hydraulic-fracture length is, the earlier the BFP ends. Therefore, this flow period will occur when the BFP of the shorter hydraulic fracture ends and the longer hydraulic fracture still lies in the BFP. The pseudo-pressure derivative curve appears as a straight line with the slope in the range of 1/4–1/2.

(5) Linear flow period (LFP): When the BFP of each hydraulic fracture ends, the LFP will appear. In this period, the linear flow perpendicular to the hydraulic-facture surfaces in the reservoir dominates in the area near the wellbore. The pseudo-pressure derivative curve appears as a straight line with the slope being 1/2.

(6) Transitional flow after linear flow period (TFALFP): When the pressure wave continues to travel in the reservoir, the interference between the pressure waves from different hydraulic fractures will take place, which makes the slope of the pseudo-pressure derivative curve be greater than 1/2. It should be noted that the stress sensitivity begins to obviously affect the type curves in this period. As the magnitude of the γ_D increases, the pseudo-pressure and its derivative curves are shifted up.

(7) Pseudo-radial flow period (PRFP): Compared with the previous flow period (i.e., the TFALFP), much slower growth of the pseudo-pressure drop was observed. The pseudo-pressure derivative without the effect of the stress sensitivity (i.e., $\gamma_D = 0$) keeps a value of 0.5, while the magnitude of the pseudo-pressure derivative with the effect of the stress sensitivity (i.e., $\gamma_D > 0$) increases with the time. Furthermore, the pseudo-pressure and its derivative increase with increasing the value of the γ_D at a fixed time. Therefore, if the stress sensitivity of the gas reservoir is stronger, the larger pressure drop is needed to remain the constant rate of produced well.

In order to quantify the effect of the stress sensitivity on the pseudo-pressure, the relative difference between the pseudo-pressures with and without the effect of the stress sensitivity is introduced as follows

$$\delta = \frac{|\psi_{wDl} - \psi_{wDnl}|}{\psi_{wDl}} \times 100\%, \tag{39}$$

where ψ_{wDnl} and ψ_{wDl} are the bottom-hole pseudo-pressures with and without the effect of the stress sensitivity, respectively.

Figure 8 shows the effect of the dimensionless permeability modulus (γ_D) on the relative difference between the pseudo-pressures with and without the effect of the stress sensitivity. With the increase of the time, the relative difference δ first increases very slowly and finally increases rapidly. Considering the flow regimes, it was found that the impact of the stress sensitivity on the pseudo-pressure was negligibly small during and before the LFP, while after the LFP, this impact became more and more obvious with the increase of the time. Furthermore, the δ became larger with increasing the magnitude of the γ_D at a fixed time.

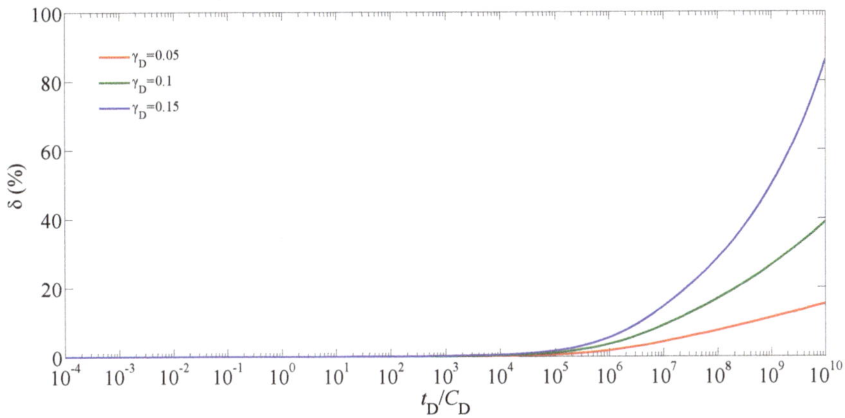

Figure 8. Effect of the γ_D on the relative difference between the pseudo-pressures with and without the effect of the stress sensitivity ($n = 4$, $L_{f1} = L_{f2} = X_f = 20$ m, $L_{f3} = L_{f4} = Y_f = 80$ m, $\theta = 30°$, $C_{fD} = 5 \times 10^4$, $S_f = 10^{-3}$, $C_D = 0.1$).

5.2. Sensitivity Analysis

Besides the dimensionless permeability modulus (γ_D), other parameters may have effects on the type curve and the relative difference between the pseudo-pressures with and without the effect of the stress sensitivity. Therefore, it is critical to conduct the sensitivity analysis of relevant parameters.

Figures 9 and 10 show the effects of the fracture conductivity coefficient (C_{fD}) on the type curve and the relative difference between the pseudo-pressures with and without the effect of the stress sensitivity, respectively. It is obvious that the impact of the C_{fD} on the type curve takes place in the early periods (i.e., from the TFAWSP to the LFP), where the values of the pseudo-pressure and its derivative increase with the decrease of the magnitude of the C_{fD}. Furthermore, it was found from Figure 9 that as the value of the C_{fD} increases, the duration of the BFP becomes shorter and the start time of the LFP becomes earlier. As shown in Figure 10, the δ was little affected by the C_{fD} in all periods, and this was because the main impact of the C_{fD} on the pseudo-pressure occurred in the early periods, but the effect of the stress sensitivity on the pseudo-pressure was negligibly small in the early periods.

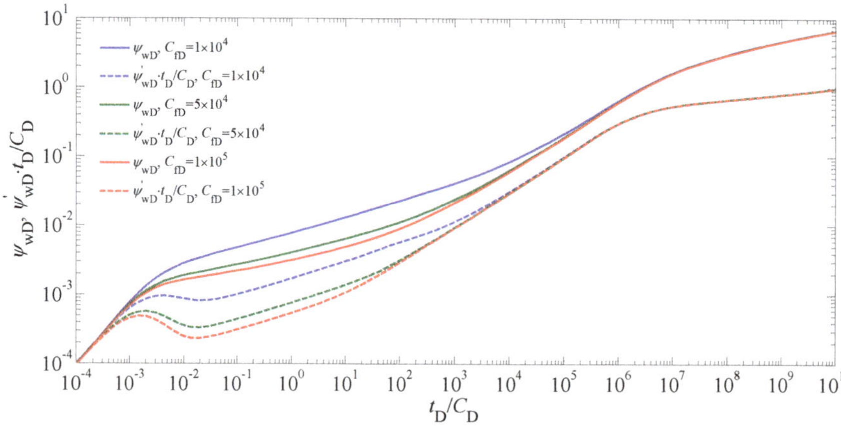

Figure 9. Type curves for the bottom-hole pseudo-pressure of a vertical fractured well with multiple hydraulic fractures in stress-sensitive gas reservoirs with different values of the C_{fD} ($n = 4$, $L_{f1} = L_{f2} = X_f = 50$ m, $L_{f3} = L_{f4} = Y_f = 50$ m, $\theta = 90°$, $S_f = 10^{-3}$, $C_D = 0.1$, $\gamma_D = 0.1$).

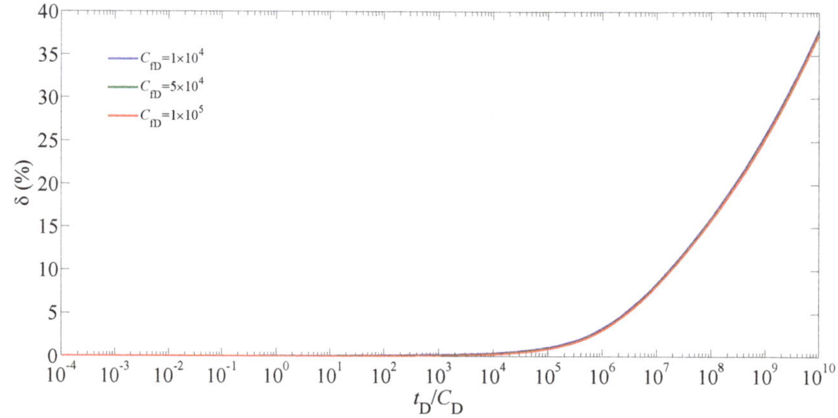

Figure 10. Effect of the C_{fD} on the relative difference between the pseudo-pressures with and without the effect of the stress sensitivity ($n = 4$, $L_{f1} = L_{f2} = X_f = 50$ m, $L_{f3} = L_{f4} = Y_f = 50$ m, $\theta = 90°$, $S_f = 10^{-3}$, $C_D = 0.1$, $\gamma_D = 0.1$).

Figures 11 and 12 show the effects of the length of the hydraulic fractures (L_f) on the type curve and the relative difference between the pseudo-pressures with and without the effect of the stress sensitivity, respectively. It is seen from Figure 11 that the L_f has an important effect on the magnitudes of the pseudo-pressure and its derivative in the intermediate and late flow periods (i.e., from the BFP to the PRFP). Decreasing the L_f reduced the duration of the BFP and resulted in the increase of the pseudo-pressure and its derivative during and after the BFP. Furthermore, it was found from Figure 12 that the δ increased with decreasing the magnitude of the L_f in the late flow period, indicating that the pseudo-pressure obtained by the conventional model without the effect of the stress sensitivity will result in a much bigger error when the length of the hydraulic fractures becomes shorter.

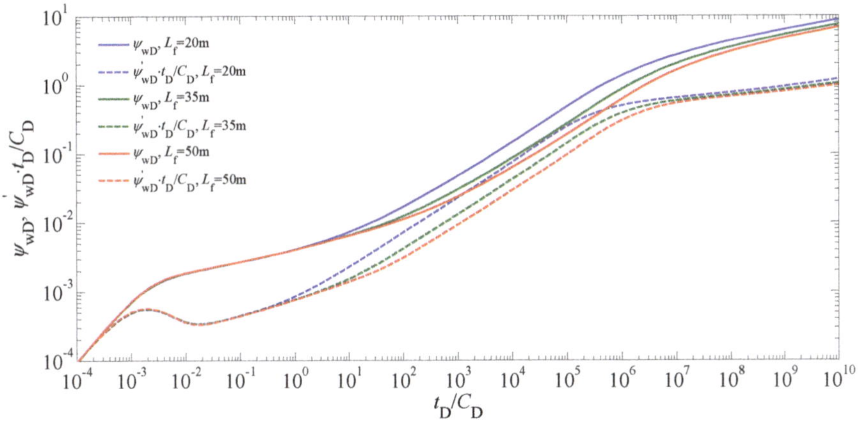

Figure 11. Type curves for the bottom-hole pseudo-pressure of a vertical fractured well with multiple hydraulic fractures in stress-sensitive gas reservoirs with different values of the L_f ($n = 4$, $L_f = L_{f1} = L_{f2} = L_{f3} = L_{f4} = X_f = Y_f$, $\theta = 90°$, $C_{fD} = 5 \times 10^4$, $S_f = 10^{-3}$, $C_D = 0.1$, $\gamma_D = 0.1$).

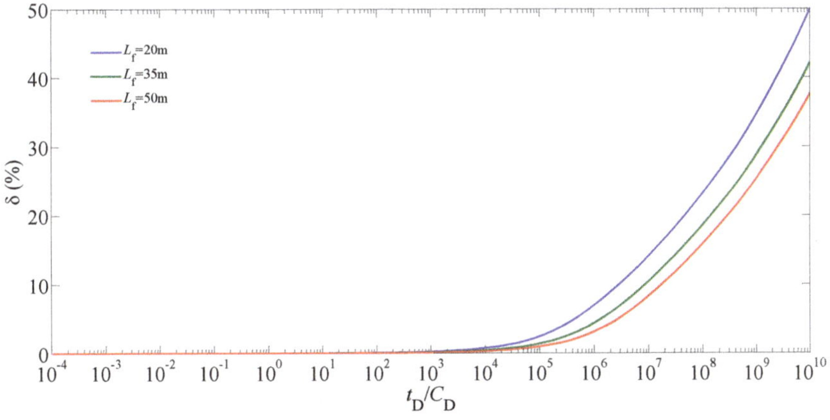

Figure 12. Effect of the L_f on the relative difference between the pseudo-pressures with and without the effect of the stress sensitivity ($n = 4$, $L_f = L_{f1} = L_{f2} = L_{f3} = L_{f4} = X_f = Y_f$, $\theta = 90°$, $C_{fD} = 5 \times 10^4$, $S_f = 10^{-3}$, $C_D = 0.1$, $\gamma_D = 0.1$).

Figures 13 and 14 show the effects of the angle between the two adjacent hydraulic fractures (θ) on the type curve and the relative difference between the pseudo-pressures with and without the effect of the stress sensitivity, respectively. As shown in Figure 13, with the decrease of the θ, the start time of the TFALFP becomes earlier, and the magnitudes of the pseudo-pressure and its derivative in the TFALFP increase. This is because decreasing the θ results in the earlier and stronger interference between the hydraulic fractures, which makes the pressure drop increase to remain the constant rate of the produced well. Whereas the impact of the θ on the type curve is relatively unobvious in the PRFP.

Furthermore, it is seen from Figure 14 that the δ increases with decreasing the magnitude of the θ during and after the TFALFP, indicating that the effect of the stress sensitivity on the pseudo-pressure becomes greater when the θ decreases.

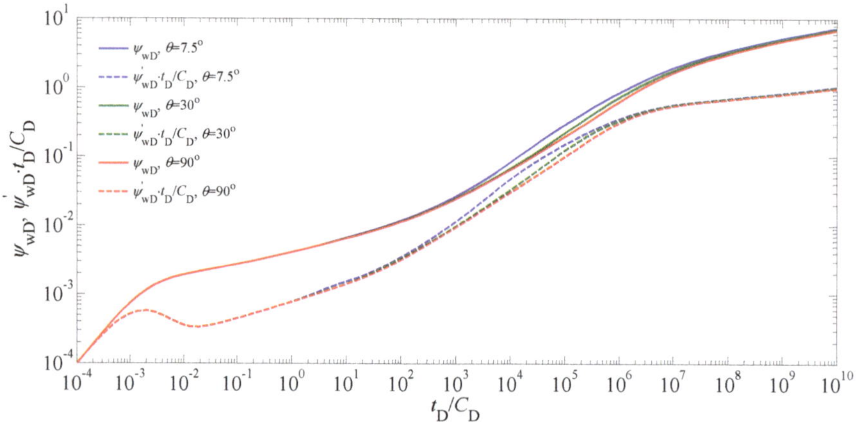

Figure 13. Type curves for the bottom-hole pseudo-pressure of a vertical fractured well with multiple hydraulic fractures in stress-sensitive gas reservoirs with different values of the θ ($n = 4$, $L_{f1} = L_{f2} = X_f = 50$ m, $L_{f3} = L_{f4} = Y_f = 50$ m, $C_{fD} = 5 \times 10^4$, $S_f = 10^{-3}$, $C_D = 0.1$, $\gamma_D = 0.1$).

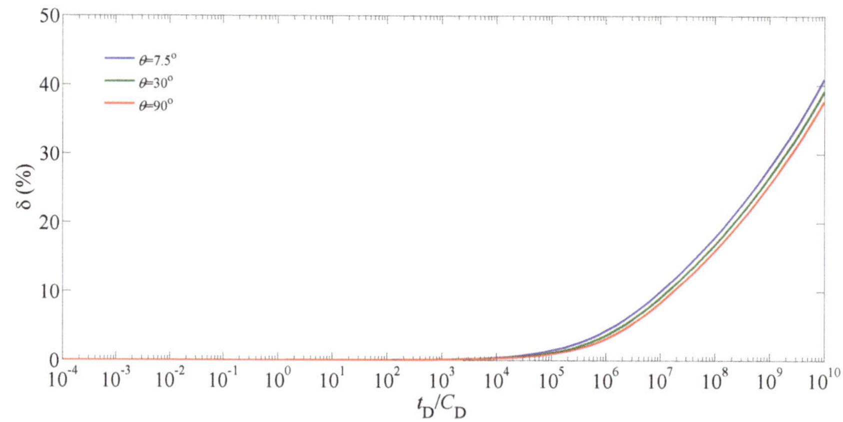

Figure 14. Effect of the θ on the relative difference between the pseudo-pressures with and without the effect of the stress sensitivity ($n = 4$, $L_{f1} = L_{f2} = X_f = 50$ m, $L_{f3} = L_{f4} = Y_f = 50$ m, $C_{fD} = 5 \times 10^4$, $S_f = 10^{-3}$, $C_D = 0.1$, $\gamma_D = 0.1$).

Introducing the coefficient of $\beta = |X_f - Y_f|/(X_f + Y_f)$ to quantify the difference of the hydraulic-fracture lengths, the effects of β on the type curve, and the relative differences between the pseudo-pressures with and without the effect of the stress sensitivity are shown in Figures 15 and 16, respectively. It is obvious that the difference of the hydraulic-fracture lengths becomes larger as the magnitude of the β increases. The range of the β is from zero to one. As shown in Figure 15, with the decrease of the β, the end of the BFP takes place later, the duration of the TFABFP becomes shorter, and the start time of the LFP occurs earlier. The reason for the existence of the TFABFP is that the difference of the hydraulic-fracture lengths leads to the inconsistency of the end times for the BFP for each hydraulic fracture. If β is equal to zero, the TFABFP may not even appear. Furthermore, the difference of the hydraulic-fracture lengths makes the pseudo-pressure and its derivative increase in the TFABFP. It is interesting to find in Figure 15 that during the TFALFP, the pseudo-pressure and its derivative increase as the β decreases. This may be because that for the distribution of the hydraulic fractures shown in Figure 6, the interference between hydraulic fractures enhances with the decrease of the β in the TFALFP. As shown in Figure 16, the β has an effect on the error caused by neglecting the stress sensitivity during and after the TFALFP. Decreasing the β results in the increase of the error.

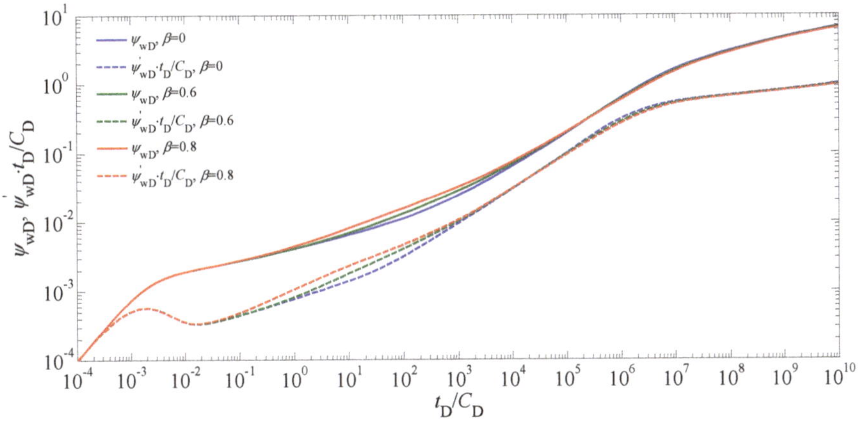

Figure 15. Type curves for the bottom-hole pseudo-pressure of a vertical fractured well with multiple hydraulic fractures in stress-sensitive gas reservoirs with different values of the β ($n = 4$, $L_{f1} = L_{f2} = X_f$, $L_{f3} = L_{f4} = Y_f$, $X_f + Y_f = 100$ m, $\theta = 90°$, $C_{fD} = 5 \times 10^4$, $S_f = 10^{-3}$, $C_D = 0.1$, $\gamma_D = 0.1$).

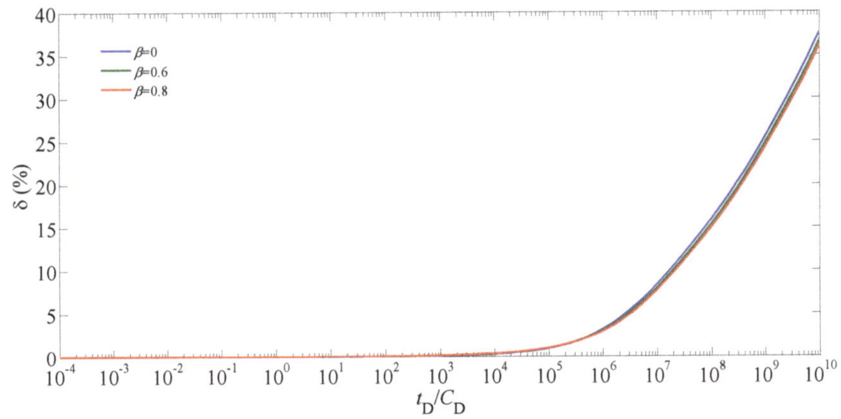

Figure 16. Effect of the β on the relative difference between the pseudo-pressures with and without the effect of the stress sensitivity ($n = 4$, $L_{f1} = L_{f2} = X_f$, $L_{f3} = L_{f4} = Y_f$, $X_f + Y_f = 100$ m, $\theta = 90°$, $C_{fD} = 5 \times 10^4$, $S_f = 10^{-3}$, $C_D = 0.1$, $\gamma_D = 0.1$).

Figures 17 and 18 show the effects of the hydraulic-fracture number (n) on the type curve and the relative difference between the pseudo-pressures with and without the effect of the stress sensitivity, respectively. The distribution of the hydraulic fractures was assumed to be uniform (i.e., the angle between the two arbitrary adjacent hydraulic fractures was equal). As shown in Figure 17, it is obvious that the n affects the pseudo-pressure behavior in all periods except for the WSP. Increasing the n leads to the decrease of the pseudo-pressure and its derivative. Due to the increase of the n, the angle between the two adjacent hydraulic fractures was reduced, and thus the interference between hydraulic fractures occurred earlier and the effect of this interference on the pseudo-pressure was more significant. Figure 18 indicates that a smaller value of the n leads to a bigger error caused by neglecting the stress sensitivity during and after the TFALFP.

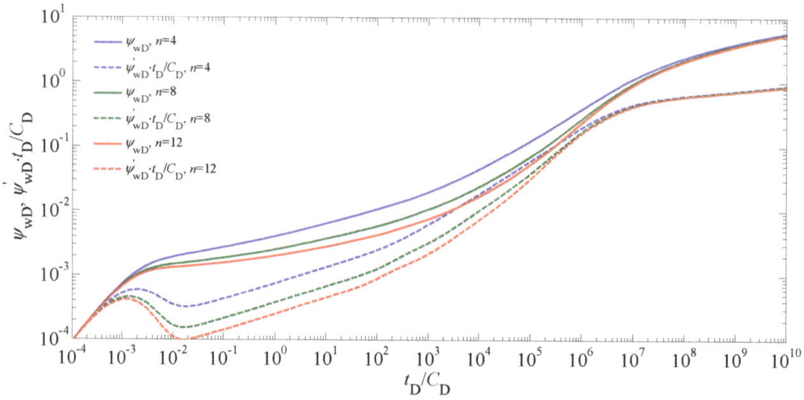

Figure 17. Type curves for the bottom-hole pseudo-pressure of a vertical fractured well with multiple hydraulic fractures in stress-sensitive gas reservoirs with different values of the n ($L_{f1} = L_{f2} = X_f =$ 80 m, $L_{f3} = L_{f4} = Y_f = 80$ m, $\theta = 360°/n$, $C_{fD} = 5 \times 10^4$, $S_f = 10^{-3}$, $C_D = 0.1$, $\gamma_D = 0.1$).

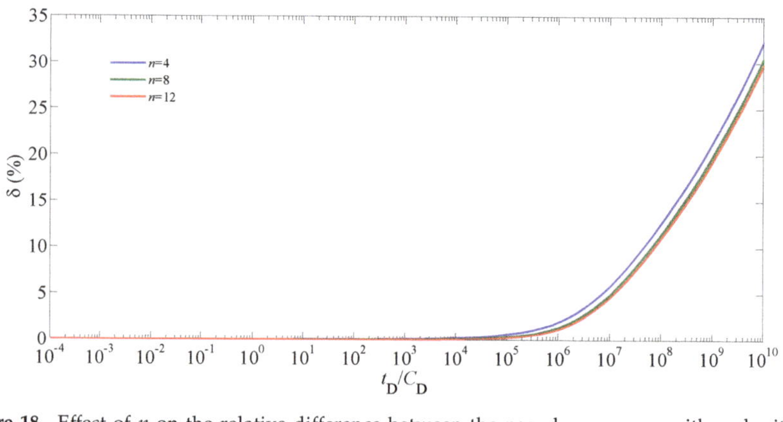

Figure 18. Effect of n on the relative difference between the pseudo-pressures with and without the effect of the stress sensitivity ($L_{f1} = L_{f2} = X_f = 80$ m, $L_{f3} = L_{f4} = Y_f = 80$ m, $\theta = 360°/n$, $C_{fD} = 5 \times 10^4$, $S_f = 10^{-3}$, $C_D = 0.1$, $\gamma_D = 0.1$).

6. Conclusions

We have established a non-linear model of a vertical fractured well with multiple hydraulic fractures in gas reservoirs by incorporating the pressure-dependent permeability. Both transformed pseudo-pressure and perturbation techniques were employed to linearize the proposed model. Superposition principle and numerical discrete methods were used to obtain the semi-analytical solution. Type curves for pseudo-pressure were plotted and discussed in detail. The influence of relevant parameters on the type curve and the relative difference between the pseudo-pressures with and without the effect of the stress sensitivity were analyzed. Some main conclusions are listed as follows:

(1) The type curve of a vertical fractured well with multiple hydraulic fractures was identified by seven possible flow regions. Compared with the conventional model of a vertical fractured well with two symmetrical hydraulic-fracture wings, two transitional flow regimes (i.e., the TFABFP and

TFALFP) were observed, which were caused by the difference of the hydraulic-fracture lengths and the interference between hydraulic fractures, respectively.

(2) As the time increased, the pressure drops increased, which made the permeability decrease. The impact of the stress sensitivity on the pseudo-pressure increased with the increase of the time. Furthermore, the influence of the stress sensitivity on the pseudo-pressure was negligibly small in the early and intermediate flow periods but becomes very significant in the late flow period.

(3) Some relevant parameters, such as dimensionless permeability modulus, fracture conductivity coefficient, hydraulic-fracture length, angle between the two adjacent hydraulic fractures, the difference of the hydraulic-fracture lengths, and hydraulic-fracture number, not only affected the type curve, but also have an influence of the error caused by neglecting the stress sensitivity.

Author Contributions: Data curation, Z.S.; Formal analysis, P.G.; Funding acquisition, J.R.; Investigation, Z.S.; Methodology, P.G.; Project administration, P.G.; Visualization, H.C.; Writing—original draft, Z.S.; Writing—review and editing, C.P.

Funding: We acknowledge the support provided by the Open Fund of the State Key Laboratory of Oil and Gas Reservoir Geology and Exploitation (PLN201627) and the Scientific Research Starting Project of SWPU (No. 2017QHZ031).

Conflicts of Interest: The authors declare no conflict of interest.

Appendix A. Symbol Description

Nomenclature

C	wellbore storage coefficient, m^3/Pa
C_{fD}	fracture conductivity coefficient
c_{ft}	total compressibility of hydraulic fracture, Pa^{-1}
c_t	total compressibility of gas reservoir, Pa^{-1}
h	reservoir thickness, m
k	reservoir permeability, m^2
k_{fi}	initial fracture permeability, m^2
k_i	initial reservoir permeability, m^2
L_{fj}	length of the jth hydraulic fracture, m
n	hydraulic-fracture number
p	reservoir pressure, Pa
p_f	fracture pressure, Pa
p_i	initial reservoir pressure, Pa
p_{sc}	pressure at standard condition, Pa
q	flow rate of point source, m^3/s
q_f	flow-rate density, m^2/s
Q_i	production rate of the ith hydraulic fracture, m^3/s
Q_{sc}	well production rate, m^3/s
r	radial distance, $r = \sqrt{x^2 + y^2}$, m
r_w	wellbore radius, m
s	Laplace transform variable
S_f	skin factor
t	time, s
T	reservoir temperature, K
T_{sc}	temperature at standard condition, K
w_f	fracture width, m
x_i, y_i	x_i- and y_i-coordinates, m
Z	Z-factor of gas
μ	gas viscosity, $Pa \cdot s$
γ	permeability modulus, $(Pa \cdot s)/Pa^2$
ψ	reservoir pseudo-pressure, $Pa^2/(Pa \cdot s)$
ψ_f	fracture pseudo-pressure, $Pa^2/(Pa \cdot s)$
ψ_i	initial reservoir pseudo-pressure, $Pa^2/(Pa \cdot s)$

ψ_w	bottom-hole pseudo-pressure, $Pa^2/(Pa \cdot s)$
ξ_D	dimensionless transformed reservoir pseudo-pressure
ξ_{fD}	dimensionless transformed fracture pseudo-pressure
ξ_{wD}	dimensionless transformed bottom-hole pseudo-pressure with wellbore storage and skin
ξ_{wHD}	dimensionless transformed bottom-hole pseudo-pressure without wellbore storage and skin
ϕ	reservoir porosity
ϕ_f	fracture porosity
θ	polar angle, degree
θ_{fj}	polar angle of the jth hydraulic fracture, degree
ΔL_{fi}	length of discrete segment of the ith hydraulic fracture, m
Subscript	
D	dimensionless
i, j	the jth discrete segment in the ith hydraulic fracture
Superscript	
$-$	Laplace space

Appendix B. Pressure Distribution in a Gas Reservoir with a Vertical Fractured Well with Multiple Hydraulic Fractures Producing at a Constant Rate

With the aid of Equation (11), Equations (7)–(10) can be rewritten as

$$\frac{\partial^2 \xi_D}{\partial r_D^2} + \frac{1}{r_D}\frac{\partial \xi_D}{\partial r_D} = \frac{1}{1-\gamma_D \xi_D}\frac{\partial \xi_D}{\partial t_D}, \tag{A1}$$

$$\xi_D(r_D, t_D = 0) = 0, \tag{A2}$$

$$\xi_D(r_D, t_D = 0) = 0, \tag{A3}$$

$$\lim_{r_D \to 0} r_D \frac{\partial \xi_D}{\partial r_D} = -q_D(r_D = 0, t_D). \tag{A4}$$

The perturbation technique of γ_D was employed to obtain that

$$\xi_D = \xi_{D0} + \gamma_D \xi_{D1} + \gamma_D^2 \xi_{D2} + \cdots, \tag{A5}$$

$$-\frac{1}{\gamma_D}\ln[1-\gamma_D \xi_D] = \xi_D + \frac{1}{2}\gamma_D \xi_D^2 + \cdots, \tag{A6}$$

$$\frac{1}{1-\gamma_D \xi_D} = 1 + \gamma_D \xi_D + \gamma_D^2 \xi_D^2 + \cdots. \tag{A7}$$

Considering a small value of γ_D in general, the zero-order perturbation solution was accurate enough for engineering requirements. So, Equations (A1)–(A4) can be rewritten as

$$\frac{\partial^2 \xi_{D0}}{\partial r_D^2} + \frac{1}{r_D}\frac{\partial \xi_{D0}}{\partial r_D} = \frac{\partial \xi_{D0}}{\partial t_D}, \tag{A8}$$

$$\xi_{D0}(r_D, t_D = 0) = 0, \tag{A9}$$

$$\xi_{D0}(r_D \to \infty, t_D) = 0, \tag{A10}$$

$$\lim_{r_D \to 0} r_D \frac{\partial \xi_{D0}}{\partial r_D} = -q_D(r_D = 0, t_D). \tag{A11}$$

Laplace transform was employed to solve the above model and it is defined by

$$\overline{f}(s) = \int_0^\infty f(t_D) e^{-st_D} \mathrm{d}t_D, \tag{A12}$$

where f is an arbitrary variable in real space.

Taking the Laplace transform of Equations (A8)–(A11), one can obtain

$$\frac{\partial^2 \bar{\xi}_{D0}}{\partial r_D^2} + \frac{1}{r_D}\frac{\partial \bar{\xi}_{D0}}{\partial r_D} = s\bar{\xi}_{D0}, \tag{A13}$$

$$\bar{\xi}_{D0}(r_D \to \infty, s) = 0, \tag{A14}$$

$$\lim_{r_D \to 0} r_D \frac{\partial \bar{\xi}_{D0}}{\partial r_D} = -\bar{q}_D(r_D = 0, s). \tag{A15}$$

The solution of Equations (A13)–(A15) in Laplace space can be derived as

$$\bar{\xi}_{D0}(r_D, s) = \bar{q}_D(r_D = 0, s) K_0(r_D\sqrt{s}). \tag{A16}$$

According to Equation (A16), the superposition principle [39] was employed to obtain the pressure distribution in a gas reservoir with a vertical fractured well with multiple hydraulic fractures producing at a constant rate

$$\bar{\xi}_{D0}(r_D, \theta, s) = \sum_{j=1}^{n}\left[\int_0^{L_{fjD}} \bar{q}_{fD}(\alpha, \theta_{fj}, s) K_0\left(\sqrt{s}\sqrt{r_D^2 + \alpha^2 - 2r_D\alpha\cos(\theta - \theta_{fj})}\right)d\alpha\right]. \tag{A17}$$

Appendix C. Derivation of Dimensionless Model for Gas Flow within Stress-Sensitive Hydraulic Fractures

Considering the effect of the pressure-dependent permeability of hydraulic fractures, the governing equation for gas flow in the ith hydraulic fracture is

$$\frac{\partial^2 \psi_f}{\partial x_i^2} + \frac{\partial^2 \psi_f}{\partial y_i^2} + \gamma\left[\left(\frac{\partial \psi_f}{\partial x_i}\right)^2 + \left(\frac{\partial \psi_f}{\partial y_i}\right)^2\right] = e^{\gamma(\psi_i - \psi_f)}\frac{\phi_f \mu c_{ft}}{k_{fi}}\frac{\partial \psi_f}{\partial t}, \tag{A18}$$

where $0 < x_i < L_{fi}$, $-\frac{w_f}{2} < y_i < \frac{w_f}{2}$.

The volume within the hydraulic fracture was very small, and thus the compressibility of the hydraulic fracture was neglected. Then Equation (A18) was reduced to

$$\frac{\partial^2 \psi_f}{\partial x_i^2} + \frac{\partial^2 \psi_f}{\partial y_i^2} + \gamma\left[\left(\frac{\partial \psi_f}{\partial x_i}\right)^2 + \left(\frac{\partial \psi_f}{\partial y_i}\right)^2\right] = 0. \tag{A19}$$

The flow-rate density of the ith hydraulic fracture is given as

$$q_f(x_i, t) = \frac{hT_{sc}k_i e^{-\gamma(\psi_i - \psi)}}{2p_{sc}T}\left[\left.\frac{\partial \psi(x_i, y_i, t)}{\partial y_i}\right|_{y_i = \frac{w_f}{2}} - \left.\frac{\partial \psi(x_i, y_i, t)}{\partial y_i}\right|_{y_i = -\frac{w_f}{2}}\right], \tag{A20}$$

where $0 < x_i < L_{fi}$.

The interface boundary condition between the wellbore and the ith hydraulic fracture is

$$h\int_{-\frac{w_f}{2}}^{\frac{w_f}{2}} \frac{T_{sc}k_{fi}e^{-\gamma(\psi_i - \psi_f)}}{2p_{sc}T}\frac{\partial \psi_f(x_i, y_i, t)}{\partial x_i}dy_i\bigg|_{x_i = 0} = Q_i(t), \tag{A21}$$

where $Q_i(t)$ is the production rate from the ith hydraulic fracture, which is expressed as

$$Q_i(t) = \int_0^{L_{fi}} q_f(x_i, t)dx_i. \tag{A22}$$

The total production rate of the vertical fractured well is the sum of the production rate of each hydraulic fracture, and thus we can obtain

$$\sum_{i=1}^{n} Q_i(t) = Q_{\rm sc}. \tag{A23}$$

The gas velocity should be continuous at the interface between the hydraulic fracture and reservoir, and thus the boundary conditions of the fracture surface are given as

$$\frac{T_{\rm sc}k_{\rm fi}e^{-\gamma(\psi_{\rm i}-\psi_{\rm f})}}{2p_{\rm sc}T}\frac{\partial\psi_{\rm f}(x_i,y_i,t)}{\partial y_i}\bigg|_{y_i=\frac{w_{\rm f}}{2}} = \frac{T_{\rm sc}k_{\rm i}e^{-\gamma(\psi_{\rm i}-\psi)}}{2p_{\rm sc}T}\frac{\partial\psi(x_i,y_i,t)}{\partial y_i}\bigg|_{y_i=\frac{w_{\rm f}}{2}}, \tag{A24}$$

$$\frac{T_{\rm sc}k_{\rm fi}e^{-\gamma(\psi_{\rm i}-\psi_{\rm f})}}{2p_{\rm sc}T}\frac{\partial\psi_{\rm f}(x_i,y_i,t)}{\partial y_i}\bigg|_{y_i=-\frac{w_{\rm f}}{2}} = \frac{T_{\rm sc}k_{\rm i}e^{-\gamma(\psi_{\rm i}-\psi)}}{2p_{\rm sc}T}\frac{\partial\psi(x_i,y_i,t)}{\partial y_i}\bigg|_{y_i=-\frac{w_{\rm f}}{2}}. \tag{A25}$$

Introducing the following expressions [16]

$$\psi_{\rm f} = \psi_{\rm i} + \frac{1}{\gamma}\ln[1-\gamma\cdot\xi_{\rm f}], \tag{A26}$$

$$\psi = \psi_{\rm i} + \frac{1}{\gamma}\ln[1-\gamma\cdot\xi], \tag{A27}$$

Equations (A19)–(A21), (A24), and (A25) can be rewritten as

$$\frac{\partial^2\xi_{\rm f}(x_i,y_i,t)}{\partial x_i^2} + \frac{\partial^2\xi_{\rm f}(x_i,y_i,t)}{\partial y_i^2} = 0,\ \left(0<x_i<L_{\rm fi},\ -\frac{w_{\rm f}}{2}<y_i<\frac{w_{\rm f}}{2}\right), \tag{A28}$$

$$q_{\rm f}(x_i,t) = -\frac{hT_{\rm sc}k_{\rm i}}{2p_{\rm sc}T}\left[\frac{\partial\xi(x_i,y_i,t)}{\partial y_i}\bigg|_{y_i=\frac{w_{\rm f}}{2}} - \frac{\partial\xi(x_i,y_i,t)}{\partial y_i}\bigg|_{y_i=-\frac{w_{\rm f}}{2}}\right],\ (0<x_i<L_{\rm fi}), \tag{A29}$$

$$h\int_{-\frac{w_{\rm f}}{2}}^{\frac{w_{\rm f}}{2}}\frac{T_{\rm sc}k_{fi}}{2p_{\rm sc}T}\frac{\partial\xi_{\rm f}(x_i,y_i,t)}{\partial x_i}{\rm d}y_i\bigg|_{x_i=0} = -Q_i(t), \tag{A30}$$

$$\frac{T_{\rm sc}k_{\rm fi}}{2p_{\rm sc}T}\frac{\partial\xi_{\rm f}(x_i,y_i,t)}{\partial y_i}\bigg|_{y_i=\frac{w_{\rm f}}{2}} = \frac{T_{\rm sc}k_{\rm i}}{2p_{\rm sc}T}\frac{\partial\xi(x_i,y_i,t)}{\partial y_i}\bigg|_{y_i=\frac{w_{\rm f}}{2}}, \tag{A31}$$

$$\frac{T_{\rm sc}k_{\rm fi}}{2p_{\rm sc}T}\frac{\partial\xi_{\rm f}(x_i,y_i,t)}{\partial y_i}\bigg|_{y_i=-\frac{w_{\rm f}}{2}} = \frac{T_{\rm sc}k_{\rm i}}{2p_{\rm sc}T}\frac{\partial\xi(x_i,y_i,t)}{\partial y_i}\bigg|_{y_i=-\frac{w_{\rm f}}{2}}. \tag{A32}$$

The width of the hydraulic fracture was very small, and thus the pressure changes within the ith hydraulic fracture in the y_i direction can be neglected. Then the average pressure within the ith hydraulic fracture in the y_i direction is introduced as follows:

$$\xi_{\rm f}(x_i,t) = \frac{1}{w_{\rm f}}\int_{-\frac{w_{\rm f}}{2}}^{\frac{w_{\rm f}}{2}}\xi_{\rm f}(x_i,y_i,t){\rm d}y_i,\ (0<x_i<L_{\rm fi}). \tag{A33}$$

With the aid of Equations (A31)–(A33), Equation (A28) becomes

$$\frac{\partial^2\xi_{\rm f}(x_i,t)}{\partial x_i^2} + \frac{k_{\rm i}}{w_{\rm f}k_{\rm fi}}\left[\frac{\partial\xi(x_i,y_i,t)}{\partial y_i}\bigg|_{y_i=\frac{w_{\rm f}}{2}} - \frac{\partial\xi(x_i,y_i,t)}{\partial y_i}\bigg|_{y_i=-\frac{w_{\rm f}}{2}}\right] = 0,\ (0<x_i<L_{\rm fi}). \tag{A34}$$

Substituting Equation (A29) into Equation (A34) yields

$$\frac{\partial^2 \xi_f(x_i,t)}{\partial x_i^2} - \frac{2p_{sc}T}{w_f k_{fi} h T_{sc}} q_f(x_i,t) = 0, \ (0 < x_i < L_{fi}). \tag{A35}$$

According to Equation (A33), Equation (A30) is expressed as

$$hw_f \frac{T_{sc}k_{fi}}{2p_{sc}T} \frac{\partial \xi_f(x_i,t)}{\partial x_i}\bigg|_{x_i=0} = -Q_i(t). \tag{A36}$$

Introducing the dimensionless variables, Equations (A26), (A35), (A36) are rewritten as follows:

$$\psi_{fD} = -\frac{1}{\gamma_D} \ln[1 - \gamma_D \cdot \xi_{fD}], \tag{A37}$$

$$\frac{\partial^2 \xi_{fD}(x_{iD},t_D)}{\partial x_{iD}^2} - \frac{2\pi}{C_{fD}} q_{fD}(x_{iD},t_D) = 0, \ (0 < x_{iD} < L_{fiD}), \tag{A38}$$

$$\frac{\partial \xi_{fD}(x_{iD},t_D)}{\partial x_{iD}}\bigg|_{x_{iD}=0} = -\frac{2\pi}{C_{fD}} Q_{iD}(t_D). \tag{A39}$$

The dimensionless expressions of Equations (A22) and (A23) are given as

$$Q_{iD}(t_D) = \int_0^{L_{fiD}} q_{fD}(x_{iD},t_D) \mathrm{d}x_{iD}, \tag{A40}$$

$$\sum_{j=1}^{n} Q_{jD}(t_D) = 1. \tag{A41}$$

References

1. Wang, Y.; Zhang, C.; Chen, T.; Ma, X. Modeling the Nonlinear Flow for a Multiple-fractured Horizontal Well with Multiple Finite-conductivity Fractures in Triple Media Carbonate Reservoir. *J. Porous Media* **2018**, *21*, 1283–1305. [CrossRef]
2. Wang, Y.; Yi, X. Flow Modeling of Well Test Analysis for A Multiple-fractured Horizontal Well in Triple Media Carbonate Reservoir. *Int. J. Nonlinear Sci. Numer. Simul.* **2018**, *19*, 439–457. [CrossRef]
3. Wang, Y.; Yi, X. Transient Pressure Behavior of a Fractured Vertical Well with A Finite-conductivity Fracture in Triple Media Carbonate Reservoir. *J. Porous Media* **2017**, *20*, 707–722. [CrossRef]
4. Ren, J.; Guo, P. A Novel Semi-analytical Model for Finite-conductivity Multiple Fractured Horizontal Wells in Shale Gas Reservoirs. *J. Nat. Gas Sci. Eng.* **2015**, *24*, 35–51. [CrossRef]
5. Ren, J.; Guo, P. Anomalous Diffusion Performance of Multiple Fractured Horizontal Wells in Shale Gas Reservoirs. *J. Nat. Gas Sci. Eng.* **2015**, *26*, 642–651. [CrossRef]
6. Ren, J.; Guo, P. Nonlinear Seepage Model for Multiple Fractured Horizontal Wells with the Effect of the Quadratic Gradient Term. *J. Porous Media* **2018**, *21*, 223–239. [CrossRef]
7. Ren, J.; Guo, P. A General Analytical Method for Transient Flow Rate with the Stress-sensitive Effect. *J. Hydrol.* **2018**, *565*, 262–275. [CrossRef]
8. Ren, J.; Guo, P. Analytical Method for Transient Flow Rate with the Effect of the Quadratic Gradient Term. *J. Pet. Sci. Eng.* **2018**, *162*, 774–784. [CrossRef]
9. Ren, J.; Guo, P.; Peng, S.; Ma, Z. Performance of Multi-stage Fractured Horizontal Wells with Stimulated Reservoir Volume in Tight Gas Reservoirs Considering Anomalous Diffusion. *Environ. Earth Sci.* **2018**, *77*, 768. [CrossRef]
10. Ren, J.; Guo, P.; Guo, Z.; Wang, Z. A Lattice Boltzmann Model for simulating Gas Flow in Kerogen Pores. *Transp. Porous Med.* **2015**, *106*, 285–301. [CrossRef]
11. Ren, J.; Guo, P.; Peng, S.; Yang, C. Investigation on Permeability of Shale Matrix using the Lattice Boltzmann Method. *J. Nat. Gas Sci. Eng.* **2016**, *29*, 169–175. [CrossRef]
12. Ren, J.; Guo, P.; Guo, Z. Rectangular Lattice Boltzmann Equation for Gaseous Microscale Flow. *Adv. Appl. Math. Mech.* **2016**, *8*, 306–330. [CrossRef]

13. Ren, J.; Guo, P. Lattice Boltzmann Simulation of Steady Flow in a Semi-Elliptical Cavity. *Commun. Comput. Phys.* **2017**, *21*, 692–717. [CrossRef]
14. Ren, J.; Zheng, Q.; Guo, P.; Peng, S.; Wang, Z.; Du, J. Pore-scale Lattice Boltzmann Simulation of Two-component Shale Gas Flow. *J. Nat. Gas Sci. Eng.* **2019**, *61*, 46–70. [CrossRef]
15. Ren, J.; Zheng, Q.; Guo, P.; Zhao, C. Lattice Boltzmann Model for Gas Flow through Tight Porous Media with Multiple Mechanisms. *Entropy* **2019**, *21*, 133. [CrossRef]
16. Pedrosa, O.A. *Pressure Transient Response in Stress-Sensitive Formations*; SPE California Regional Meeting: Oakland, CA, USA, 1986.
17. Zhang, Z.; He, S.; Liu, G.; Guo, X.; Mo, S. Pressure Buildup Behavior of Vertically Fractured Wells with Stress-sensitive Conductivity. *J. Pet. Sci. Eng.* **2014**, *122*, 48–55. [CrossRef]
18. Wang, J.; Jia, A.; Wei, Y.; Qi, Y. Approximate Semi-analytical Modeling of Transient Behavior of Horizontal Well Intercepted by Multiple Pressure-dependent Conductivity Fractures in Pressure-sensitive Reservoir. *J. Pet. Sci. Eng.* **2017**, *153*, 157–177. [CrossRef]
19. Gringarten, A.C.; Ramey, H.J.; Raghavan, J.R. Unsteady-state Pressure Distributions Created by a Well with a Single Infinite-conductivity Vertical Fracture. *SPE J.* **1974**, *14*, 347–360. [CrossRef]
20. Cinco-Ley, H.; Samaniego, F.; Dominguez, N. Transient Pressure Behavior for a Well with a Finite-conductivity Vertical Fracture. *SPE J.* **1978**, *18*, 253–264.
21. Bennett, C.O.; Reynolds, A.C.; Raghavan, R.; Elbel, J.L. Performance of Finite-conductivity, Vertically Fractured Wells in Single-layer Reservoirs. *SPE Form. Eval.* **1986**, *1*, 399–412. [CrossRef]
22. Bennett, C.O.; Rosato, N.D.; Reynolds, A.C.; Raghavan, R. Influence of Fracture Heterogeneity and Wing Length on the Response of Vertically Fractured Wells. *SPE J.* **1983**, *23*, 219–230. [CrossRef]
23. Rodriguez, F.; Cinco-Ley, H.; Samaniego, V.F. Evaluation of Fracture Asymmetry of Finite-conductivity Fractured Wells. *SPE Prod. Eval.* **1992**, *7*, 233–239. [CrossRef]
24. Berumen, S.; Tiab, D.; Rodriguez, F. Constant Rate Solutions for a Fractured Well with an Asymmetric Fracture. *J. Pet. Sci. Eng.* **2000**, *25*, 49–58. [CrossRef]
25. Tiab, D.; Lu, J.; Nguyen, H.; Owayed, J. Evaluation of Fracture Asymmetry of Finite-conductivity Fractured Wells. *J. Energy Resour. Technol.* **2010**, *132*, 1–7. [CrossRef]
26. Wang, L.; Wang, X.; Li, J.; Wang, J. Simulation of Pressure Transient Behavior for Asymmetrically Finite-conductivity Fractured Wells in Coal Reservoirs. *Transp. Porous Media* **2013**, *97*, 353–372. [CrossRef]
27. Wang, L.; Wang, X. Type Curves Analysis for Asymmetrically Fractured Wells. *J. Energy Resour. Technol.* **2014**, *136*, 023101. [CrossRef]
28. Fisher, M.K.; Wright, C.A.; Davidson, B.M.; Goodwin, A.K.; Fielder, E.O.; Buckler, W.S.; Steinsberger, N.P. Integrating Fracture Mapping Technologies to Improve Stimulations in the Barnett Shale. *SPE Prod. Facil.* **2005**, *20*, 85–93. [CrossRef]
29. Mayerhofer, M.J.; Lolon, E.P.; Youngblood, J.E.; Heinze, J.R. *Integration of Microseismic-Fracture-Mapping Results with Numerical Fracture Network Production Modeling in the Barnett Shale*; SPE Annual Technical Conference and Exhibition: San Antonio, TX, USA, 2006.
30. Fan, Z.Q.; Jin, Z.H.; Johnson, S.E. Oil-gas Transformation Induced Subcritical Crack Propagation and Coalescence in Petroleum Source Rocks. *Int. J. Fract.* **2014**, *185*, 187–194. [CrossRef]
31. Ren, J.; Guo, P. Nonlinear Flow Model of Multiple Fractured Horizontal Wells with Stimulated Reservoir Volume including the Quadratic Gradient Term. *J. Hydrol.* **2017**, *554*, 155–172. [CrossRef]
32. Shaoul, J.R.; Behr, A.; Mtchedlishvili, G. Developing a Tool for 3D Reservoir Simulation of Hydraulically Fractured Wells. *SPE Reservoir Eval. Eng.* **2007**, *10*, 50–59. [CrossRef]
33. Restrepo, D.P.; Tiab, D. Multiple fractures transient response. In Proceedings of the SPE Latin American and Caribbean Petroleum Engineering Conference, Cartagena de Indias, Colombia, 31 May–3 June 2009.
34. Wang, L.; Wang, X.; Luo, E.; Wang, J. Analytical Modeling of Flow Behavior for Wormholes in Naturally Fractured-vuggy Porous Media. *Transp. Porous Media* **2014**, *105*, 539–558. [CrossRef]
35. Ren, J.; Guo, P. Performance of Vertical Fractured Wells with Multiple Finite-conductivity Fractures. *J. Geophys. Eng.* **2015**, *12*, 978–987. [CrossRef]
36. Luo, W.; Tang, C. Pressure-transient Analysis of Multiwing Fractures Connected to a Vertical Wellbore. *SPE J.* **2015**, *20*, 360–367. [CrossRef]
37. Wang, H.T. Performance of Multiple Fractured Horizontal Wells in Shale Gas Reservoirs with Consideration of Multiple Mechanisms. *J. Hydrol.* **2014**, *510*, 299–312. [CrossRef]

38. Li, X.P.; Cao, L.N.; Luo, C.; Zhang, B.; Zhang, J.Q.; Tan, X.H. Characteristics of Transient Production Rate Performance of Horizontal Well in Fractured Tight Gas Reservoirs with Stress-sensitivity Effect. *J. Pet. Sci. Eng.* **2017**, *158*, 92–106. [CrossRef]
39. Ozkan, E. Performance of Horizontal Wells. Ph.D. Thesis, University of Tulsa, Tulsa, OK, USA, 1988.
40. Wan, J. Well Models of Hydraulically Fractured Horizontal Wells. Master's Thesis, Stanford University, Palo Alto, CA, USA, 1999.
41. Van Everdingen, A.F.; Hurst, W. The Application of the Laplace Transformation to Flow Problems in Reservoirs. *Trans. AIME* **1949**, *186*, 305–324. [CrossRef]
42. Stehfest, H. Numerical Inversion of Laplace Transforms. *Commun. ACM* **1970**, *13*, 47–49. [CrossRef]
43. Akanji, L.T.; Falade, G.K. Closed-Form Solution of Radial Transport of Tracers in Porous Media Influenced by Linear Drift. *Energies* **2019**, *12*, 29. [CrossRef]
44. Choo, Y.K.; Wu, C.H. Transient Pressure Behavior of Multiple-fractured Gas Wells. In Proceedings of the SPE/DOE Low Permeability Reservoirs Symposium, Denver, CO, USA, 18–19 May 1987.

applied sciences

MDPI

Article

Parametrical Non-Complex Tests to Evaluate Partial Decentralized Linear-Output Feedback Control Stabilization Conditions from Their Centralized Stabilization Counterparts

Manuel De la Sen [1,*] and Asier Ibeas [2]

1 Institute of Research and Development of Processes IIDP, University of the Basque Country, Campus of Leioa, 48940 Leioa (Bizkaia), Spain

2 Department of Telecommunications and Systems Engineering, Universitat Autònoma de Barcelona UAB, 08193 Barcelona, Spain; Asier.Ibeas@uab.cat

* Correspondence: manuel.delasen@ehu.eus

Received: 18 March 2019; Accepted: 22 April 2019; Published: 26 April 2019

Abstract: This paper formulates sufficiency-type linear-output feedback decentralized closed-loop stabilization conditions if the continuous-time linear dynamic system can be stabilized under linear output-feedback centralized stabilization. The provided tests are simple to evaluate, while they are based on the quantification of the sufficiently smallness of the parametrical error norms between the control, output, interconnection and open-loop system dynamics matrices and the corresponding control gains in the decentralized case related to the centralized counterpart. The tolerance amounts of the various parametrical matrix errors are described by the maximum allowed tolerance upper-bound of a small positive real parameter that upper-bounds the various parametrical error norms. Such a tolerance is quantified by considering the first or second powers of such a small parameter. The results are seen to be directly extendable to quantify the allowed parametrical errors that guarantee the closed-loop linear output-feedback stabilization of a current system related to its nominal counterpart. Furthermore, several simulated examples are also discussed.

Keywords: Output-feedback; centralized control; decentralized control; closed-loop stabilization

---

1. Introduction

Control systems are very important in real world applications and, therefore, they have been investigated exhaustively concerning their properties of stability, stabilization, controllability control strategies etc. See, for instance, [1–4] and references therein. Some extra constraints inherent to some systems, like solution positivity in the case of biological systems or human migrations or the needed behavior robustness against parametrical changes of disturbance actions add additional complexity to the related investigations and need the use of additional mathematical or engineering tools for the research development, [5–7]. A large variety of modeling and design tools have to be invoked and developed in the analysis depending on the concrete systems under study and their potential applications as, for instance, the presence of internal and external delays, discretization, dynamics modeling based on fractional calculus, the existence of complex systems with interconnected subsystems, [8–13], hybrid coupled continuous/digital tandems, nonlinear systems and optimization and estimation techniques [14–19] as well as robotic and fuzzy-logic based systems, [20,21]. In particular, decentralized control is a useful tool for controlling dynamic systems by cutting some links between the dynamics coupling a set of subsystems integrated in the whole system at hand. It is claimed to keep the main properties related to the use of centralized control such as stability, controllability, observability, etc. In summary, a centralized controller keeps all the information on the system and coupling links as available to the control designer while decentralized control ignores some of such information or even cuts on occasions some of coupling signals between the various subsystems integrated in the whole system at hand. It can be pointed out that the stability studies are often performed trough Lyapunov theory which requires to find a Lyapunov function (see [20,21] and some references therein). It turns out that, if the neglected couplings are strong and are not taken into account by the controller, the stabilization and other properties such as the controllability can become lost. The use of decentralized stabilization and control tools is of interest when the whole system has physically separated subsystems that require the implementation of local control actuators but the control has to be global for the whole system. An ad-hoc example provided in [2–4,19] where decentralized control is of a great design interest is the case of several coupled cascade hydroelectric power plants allocated in the same river but separated far away from each other. It has to be pointed out that the term "decentralized control" versus "centralized control" refers to the eventual cut of links of the shared information between tandems of integrated subsystems, or coupling signals between them, to be controlled rather than to the physical disposal of the controller. In other words, if the whole controller keeps and uses all the information on measurable outputs and control components design to implement the control law, such a control is considered to have a centralized nature even if its various sub-control stations are not jointly allocated. It is a common designer´s basic idea in mind for complex control designs to try to minimize the modeling designs and computational loads without significantly losing the system´s performance and its essential properties. For instance, in [8], the dynamic characteristic of a discrete-time system is given as an extended state space description in which state variables and output tracking error are integrated while they are regulated independently. The proposed robust model predictive control is much simpler than the traditional versions since the information of the upper and lower bounds of the time-varying delay are used for design purposes. On the other hand, in [9], a control law might be synthesized for a hydropower plant with six generation units working in an alternation scheme. To assess the behavior of the controlled system, a model of such a nonlinear plant is controlled by a fractional proportional/integral/derivative control device through a linearization of its set points, the fractional part being relevant in the approach on the controller derivative actions. In addition, a set of applied complex control problems are studied, for instance, in [10–16] with the aim of giving different ad hoc simplification tools to deal with the appropriate control methodologies. In particular, a decentralized control approach is proposed in [16].

In this paper, the decentralized control design versus its decentralized control counterpart, under eventual output linear feedback, are studied from the point of view of the amount of information that can be lost or omitted in terms of the total or partial knowledge of the coupled dynamics between subsystems necessary in the decentralized case to keep the closed-loop stability. The study is made by using the information on the worst-case deviation, in terms of norms, between the respective matrices of open-loop dynamics and the respective controller gains under which the closed-loop stability is kept. This paper is organized as follows. The problem statement is given in Section 2 while the main stabilization results of the paper are provided in Section 3. The proofs of some of the results of Section 3, which are technically involved, but conceptually simple, are distributed in various technical auxiliary that are given in Appendices A and B. It is claimed to give a non-complex method to test the feasibility of the implementation of decentralized control and conditions for its design, which be a fast and simpler stability test compared to Lyapunov stability theory [20,21], for instance, under a partial removal of information or physical cuts of links of coupling dynamics between the various subsystems or state, control and output components. Section 4 discusses several examples and, finally, the concluding remarks end the paper.

Notation

$$\overline{n} = \{1, 2, \cdots, n\},$$

$$\boldsymbol{R}_{+} = \{z \in \boldsymbol{R} : z > 0\} \; ; \; \boldsymbol{R}_{0+} = \{z \in \boldsymbol{R} : z \geq 0\},$$

$sp(A)$ and $\det(A)$ are the spectrum and determinant of $A \in \mathbf{R}^{n\times n}$, respectively. For $A \in \mathbf{R}^{q\times \ell}$, being in general rectangular, $\|A\|$ denotes any unspecified norm of A, $\|A\|_2$ denotes the ℓ_2 or spectral norm of a matrix A, $\rho(A)$ denotes its spectral radius, and $\|.\|_\infty$ denotes the H_∞-norm of a stable real rational transfer matrix or function, I_q denotes the qth identity matrix, and $\boldsymbol{i} = \sqrt{-1}$ is the complex imaginary unit.

Let $A \in \mathbf{R}^{n\times n}$ be symmetric. Then, $\lambda_{\max}(A)$ and $\lambda_{\min}(A)$ are, respectively, the maximum and minimum eigenvalues of A.

$M_E^{n\times n}$ is the set of Metzler matrices (any off-diagonal entry is non-negative) of nth order.

$Z^{n\times n}$ is the set of Z-matrices (any off-diagonal entry is non-positive) of nth order.

$M^{n\times n}$ is the set of M-matrices (Z-matrices which are stable or critically stable) of nth order.

Assume that $A = \left(A_{ij}\right), B = \left(B_{ij}\right) \in \mathbf{R}^{n\times n}$. Then, the notations $A \underline{>} B$, $A > B$ and $A >> B$, are, respectively, equivalent to $B \underline{<} A$, $B < A$ and $B << A$, meaning that $A_{ij} \geq B_{ij}$, $A_{ij} \geq B_{ij}$ (and $B \neq A$) and $A_{ij} > B_{ij}$; $\forall i, j \in \overline{n}$, respectively. In particular, $A \underline{>} 0$, $A > 0$ and $A >> 0$ are reworded as A is non-negative, positive and strictly positive, respectively, and $A \underline{<} 0$, $A < 0$ and $A << 0$ are reworded as A is non-positive, negative and strictly negative, respectively.

2. Problem Statement

Consider the following linear and time-invariant system under linear output-feedback centralized control:

$$\dot{x}_c(t) = A_c x_c(t) + B_c\, u_c(t);\; x_c(0) = x_{c0} \tag{1}$$

$$y_c(t) = C_c x_c(t) + D_c u_c(t) \tag{2}$$

$$u_c(t) = K_c y_c(t) \tag{3}$$

where $x_c(t) \in \mathbf{R}^n$ is the state vector; $u_c(t) \in \mathbf{R}^m$ is the centralized control vector; $y_c(t) \in \mathbf{R}^p$ is the output; A_c, B_c, C_c and D_c are the system, control, output and input–output interconnection matrices, respectively, of orders being compatible with the dimensions of the above vectors; and $K_c \in \mathbf{R}^{m\times p}$ is the control matrix. If the system runs in a decentralized control context, we have:

$$\dot{x}_d(t) = A_d x_d(t) + B_d\, u_d(t)\,;\; x_d(0) = x_{d0} \tag{4}$$

$$y_d(t) = C_d x_d(t) + D_d u_d(t) \tag{5}$$

$$u_d(t) = K_d y_d(t) \tag{6}$$

where $x_d(t) \in \mathbf{R}^n$ is the state vector; $u_d(t) \in \mathbf{R}^m$ is the centralized control vector; $y_d(t) \in \mathbf{R}^p$ is the output; A_d, B_d, C_d and D_d are the system, control, output and input–output interconnection matrices, respectively, of orders being compatible with the dimensions of the above vectors; and $K_d \in \mathbf{R}^{m\times p}$ is the control matrix.

Basically, the differences between centralized and decentralized controls are as follows:

(1) In the centralized control, all control components, or more generally, all subsystems if subsystems are considered in the model, have a complete information on the output available for feedback. This means that all control components or block-control inputs are available for controlling each state component (or each individual substate including several state components in the case of a more generic decomposition structure). Basically, the matrix K_c has a complete non-diagonal or block non-diagonal structure. In the decentralized control, the various input components or block-control inputs are not available for controlling each state component. Thus, K_d does not have a complete free design structure of its non-diagonal part and in some cases (completely decentralized disposal) its diagonal or block diagonal.

(2) In a more general context, some control or output links can be cut in the decentralized case for the sake of computational simplicity or a more economic control design. In our case, the decentralized input, output and interconnection matrices B_d, C_d and D_d can be distinct from the

centralized ones and, roughly speaking, to a have a "more diagonal" or "sparser" structure than their centralized counterparts B_c, C_c and D_c. If the parameterization of the system (or dynamics) matrix is available to the designer, then some off-diagonal block matrices of A_c could be zeroed or simply re-disposed in a more sparse structure to construct A_d.

(3) The only strictly necessary condition for the system to be subject to partially (or, respectively, fully) decentralized control is that some (or, respectively, all) of the off-diagonal entries of K_d are forced to be zero even if the system, control, output and interconnection matrices remain identical in Equations (4) and (5) with respect to Equations (1) and (2).

Assumption 1. *The system in Equations (1) and (2) is linear output-feedback stabilizable via some centralized control law (Equation (3)).*

Note that Assumption 1 does not hold if the open-loop system in Equations (1) and (2) has unstable or critically stable fixed modes that cannot be removed via linear feedback.

Proposition 1. *If Assumption 1 holds, then there exists a centralized stabilizing controller gain $K_c \in \mathbf{R}^{m\times p}$ such that the matrices $(I_m - K_cD_c)$ and $\left(I_p - D_cK_c\right)$ are non-singular, thus the closed-loop centralized control system is solvable and given by:*

$$\dot{x}_c(t) = \left(A_c + B_c(I_m - K_cD_c)^{-1}K_cC_c\right)x_c(t);\ x_c(0) = x_{c0} \tag{7}$$

$$y_c(t) = \left(I_p - D_cK_c\right)^{-1}C_cx_c(t) \tag{8}$$

and asymptotically stable for any given $x_{c0} \in \mathbf{R}^n$ under the generated control law:

$$u_c(t) = (I_m - K_cD_c)^{-1}K_cC_cx_c(t) \tag{9}$$

that is, the polynomial $p(s) = \det\left(sI_n - A_c - B_c(I_m - K_cD_c)^{-1}K_cC_c\right)$ is Hurwitz.

Proof. The replacement of Equation (3) into Equations (1) and (2) yields Equation (7)–(9). Since the (1) and (2) is linear output stabilizable, a stabilizing controller gain $K_c \in \mathbf{R}^{m\times p}$ has to exist such that (7)–(9) are solvable and the closed-loop dynamics is stable. □

Assumption 2. *The system in Equations (4) and (5) is linear output-feedback stabilizable via some decentralized control law (Equation (6)).*

In the same way as Proposition 1, we get the following result:

Proposition 2. *If Assumption 2 holds, then there exists a decentralized stabilizing controller gain $K_d \in \mathbf{R}^{m\times p}$ such that the matrices $(I_m - K_dD_d)$ and $\left(I_p - D_dK_d\right)$ are non-singular, thus the closed-loop decentralized control system is solvable and given by:*

$$\dot{x}_d(t) = \left(A_d + B_d(I_m - K_dD_d)^{-1}K_dC_d\right)x_d(t);\ x_d(0) = x_{d0} \tag{10}$$

$$y_d(t) = \left(I_p - D_dK_d\right)^{-1}C_dx_d(t) \tag{11}$$

and asymptotically stable for any given $x_{d0} \in \mathbf{R}^n$ under the control law:

$$u_d(t) = (I_m - K_dD_d)^{-1}K_dC_dx_d(t) \tag{12}$$

that is, the polynomial $p(s) = \det\left(sI_n - A_d - B_d(I_m - K_dD_d)^{-1}K_dC_d\right)$ is Hurwitz.

Proposition 3. *Assume that $A_d = A_c$, $B_d = B_c$, $C_d = C_c$ and $D_d = D_c$, and that the system in Equations (1) and (2) is not linear output-feedback stabilizable via some centralized control law (Equation (3)). Then, it is not stabilizable under any linear output-feedback decentralized control (Equation (6)) either.*

Proof. Obviously, if there is no completely free-design matrix K_c that stabilizes Equations (1) and (2), then there is no K_d with at least a forced zero off-diagonal entry that stabilizes it since K_d has extra design constraints related to K_c. □

It can be pointed out that decentralized control has also been proved to be useful in applications. For instance, an integral-based event-triggered asymptotic stabilization of a continuous-time linear system is studied in [17] by considering actuator saturation and observer-based output feedback are considered. In the proposed scheme, the sensors and actuators are implemented in a decentralized manner and a type of Zeno-free decentralized integral-based event condition is designed to guarantee the asymptotic stability of the closed-loop systems. On the other hand, two decentralized fuzzy logic-based control schemes with a high-penetration non-storage wind-diesel system are studied in [18] for small power system with high-penetration wind farms. In addition, several examples concerning decentralized control are described in [4] to illustrate the theoretical design analysis. A typical described case is that of tandems of electrical power system with a tandem disposal on the same river which are not physically nearly allocated. The next section discusses some simple sufficiency-type conditions which ensure that, provided that the system is stabilizable under linear output-feedback centralized control, it is also stabilizable under decentralized control in two cases: (a) the system matrix remains identical but the other parameterization matrices can eventually vary; and (b) the system matrix can vary as well in the decentralized case with respect to the centralized one. A result elated to the maintenance of the stability of a matrix under an additive matrix perturbation is summarized through a set of sufficiency-type conditions simple to test in Theorem A1. Theorem A2 proves sufficiency-type for the stability of the matrix function $A(t) = A_0 + \widetilde{A}(t)$ with A_0 stable and $\widetilde{A}(t)$ being time-varying. Appendix B includes calculations and auxiliary results to quantify the tolerance to cut some dynamics links between subsystems, state components or control centers or components while keeping the closed-loop stability of the whole coupled system. The results of Appendices A and B are used in the proofs of the main results in the next section.

3. Main Results

The first set of technical results which follow are concerned with centralized and decentralized control stabilizability.

Assertion 1. *A necessary and sufficient condition for the system to be linear state-stabilizable via some centralized control law is that $rank[sI_n - A_c\, B_c] = n$ for all $Res \geq 0$.*

Proof. Assume that $rank[sI_n - A_c\, B_c] = n_1 < n$ for some $Res \geq 0$. Then, there is some Laplace transform $\left[\hat{x}^T(s)\,,\,\hat{u}^T(s)\right]^T = Lap\left[x^T(t)\,,\,u^T(t)\right]^T \neq 0$ such that $[sI_n - A_c\, B_c]\left[\hat{x}^T(s)\,,\,\hat{u}^T(s)\right]^T = 0$ for some $Res \geq 0$ and $[sI_n - A_c\, B_c]\begin{bmatrix} I_n \\ K_c \end{bmatrix}\hat{x}(s) = 0$ for any $K_c \in \mathbf{R}^{n\times n}$ and some $\hat{x}(s) \neq 0$ with $Res \geq 0$ since $rank\left([sI_n - A_c\, B_c]\begin{bmatrix} I_n \\ K_c \end{bmatrix}\right) \leq \min\left(rank[sI_n - A_c\, B_c]\,,\, rank\begin{bmatrix} I_n \\ K_c \end{bmatrix}\right) \leq \min(n_1\,,\, n) \leq n_1 < n$ for some $Res \geq 0$. Therefore, the closed-loop system has some unstable or critically stable solution for any given (centralized) control gain. This proves the necessary part. Sufficiency follows since, if $rank[sI_n - A_c\, B_c] = n$, then $\hat{x}(s) \equiv 0$ for all $Res \geq 0$ and some $K_c \in \mathbf{R}^{n\times n}$ which can be found so that $\min\left(rank[sI_n - A_c\, B_c]\,,\, rank\begin{bmatrix} I_n \\ K_c \end{bmatrix}\right) = \min(n, n) = n$. □

Assertion 1. is a particular adapted ad-hoc test for stabilizability of the celebrated Popov–Belevitch –Hautus rank controllability test [6]. Note that, if there exist unstable or critically stable fixed modes (i.e., those present in the open-loop system that cannot be removed via feedback control), then neither centralized nor decentralized stabilizing control laws can be synthesized. Note that the stabilizability rank test of Assumption 1 can only be evaluated for the critically and unstable eigenvalues of A_c instead for all the open right-hand complex plane. In all the remaining points of such a plane, the test always gives a full rank of the tested matrix. The parallel controllability test should always be applied in the same matrix to any eigenvalues of A_c.

Assertion 2. *A necessary condition for the system to be linear state-stabilizable via some partially or totally decentralized control is that it be stabilizable via centralized control (i.e., Assertion 1 holds).*

Proof. It is obvious from Assertion 1 since any gain K_d used for centralized or decentralized is sparser than a centralized gain counterpart so that the proof follows from Assertion 1. □

Assertion 3. *A necessary condition for the system to be linear output-stabilizable via some partially or totally decentralized control is that it be stabilizable via centralized control (i.e., that Assertion 1 holds).*

Proof. It is obvious from Assertions 1 and 2 that, when replacing $K_c \to (I_m - K_cD_c)^{-1}K_cC_c$ and $K_d \to (I_m - K_dD_d)^{-1}K_dC_d$ (see Equations (9) and (12)), the second replacement happens under sparser parameterizations. □

Now, consider the closed-loop system matrices from Equations (7) and (10) for the case $A = A_c = A_d$.

$$A_{cc} = A + B_c(I_m - K_cD_c)^{-1}K_cC_c; A_{dc} = A + B_d(I_m - K_dD_d)^{-1}K_dC_d$$

with its parametrical error being:

$$\widetilde{A}_{dc} = A_{cc} - A_{dc} = B_c(I_m - K_cD_c)^{-1}K_cC_c - B_d(I_m - K_dD_d)^{-1}K_dC_d$$

A first main technical result follows:

Theorem 1. *If Assumption 1 holds, assume also that $K_c \in \mathbf{R}^{m\times p}$ is a centralized linear output-feedback stabilizing controller gain such that the resulting closed-loop system matrix $A_{cc} \in \mathbf{R}^{n\times n}$ has a stability abscissa $(-\rho_{cc}) < 0$. Then, the following properties hold:*

(i) *$A_{dc} \in \mathbf{R}^{n\times n}$ is a closed-loop stability matrix under a linear output-feedback stabilizing controller gain $K_d \in \mathbf{R}^{m\times p}$ if any of the subsequent sufficiency-type conditions holds:*

(1) The H_∞-norm of $(sI_n - A_{cc})^{-1}\widetilde{A}_{dc}$ satisfies $\|(sI_n - A_{cc})^{-1}\widetilde{A}_{dc}\|_\infty < 1$,
(2) $\|\widetilde{A}_{dc}\|_2 < 1/\sup_{\omega\in\mathbf{R}_{0+}} \|(i\omega I_n - A_{cc})^{-1}\|_2$.
Other alternative sufficiency-type conditions to Conditions 1 and 2 for the stability of A_{dc} are:
(3) $\rho\left(A_{cc}^{-1}\widetilde{A}_{dc}\right) < 1$,
(4) $\|A_{cc}^{-1}\widetilde{A}_{dc}\|_2 < 1$,
(5) $\|\widetilde{A}_{dc}\|_2 < 1/\|A_{cc}^{-1}\|_2$, that is, $\lambda_{\max}\left(\widetilde{A}_{dc}^T\widetilde{A}_{dc}\right) < \lambda_{\min}\left(A_{cc}^TA_{cc}\right)$,
in the following particular cases:
(a) $A_{cc} < 0$ and $\widetilde{A}_{dc} > A_{cc}$; and
(b) $A_{cc} = \left(A_{cc_{ij}}\right) \in M_E^{n\times n}$ and $\widetilde{A}_{dc} = \left(\widetilde{A}_{dcij}\right)$ fulfills $\widetilde{A}_{dcij} \le A_{cc_{ij}}$; $\forall i, j(\ne i) \in \overline{n}$.

(ii) *Assume that Property (i) holds and that the number of inputs and outputs are identical, i.e., $p = m$, and decompose both the controller gain matrices as sums of their diagonal and off-diagonal parts leading to $K_c = K_{cd} + K_{cod}$ and $K_d = K_{dd} + K_{dod}$, thus $\widetilde{K} = K_c - K_d = (K_{cd} - K_{dd}) + (K_{cod} - K_{dod})$. Then, the system is stabilizable under partially decentralized control linear output-feedback control in the sense that Equations (4) and (5), is asymptotically stable under a control law (Equation (6)), if $K_d \in \mathbf{R}^{m\times p}$ is such that, if there* *is at least one non-diagonal zero entry in at least one of its rows in the off-diagonal controller error matrix* $\widetilde{K}_{od} = K_{cod} - K_{dod}$*. If* $\widetilde{K}_{od} = 0$*, then the system is stabilizable under decentralized control.*

Proof. Property (i) is a direct translation of the results of Theorem A1 in Appendix A to the closed-loop system matrices. Property (ii) holds if Property (i) holds with an off-diagonal controller error matrix between the centralized a decentralized controller gain that has at least one non-diagonal zero at some row (or its identically zero) so that a feedback from some crossed output to some of the inputs is not provided to the control law for closed-loop stabilization the stabilization. □

The following result follows for the time-varying case from Theorem 1 and Theorem A2:

Corollary 1. *Assume that* $A_{dc}(t)$ *and then* $\widetilde{A}_{dc}(t)$ *are everywhere piecewise-continuous time-varying. Then, Theorem 1 still holds if Condition 1 is replaced with* $\frac{k_{cc}}{\rho_{cc}} \sup_{0 \le \tau < t} \|\widetilde{A}_{dc}(\tau)\| < 1$*;* $\forall t \in \mathbf{R}_{0+}$ *with* $k_{cc} \ge 1$ *and* $\rho_{cc} > 0$ *being real constants such that* $e^{-A_{cc}t} \le k_{cc}e^{-\rho_{cc}t}$*;* $\forall t \in \mathbf{R}_{0+}$*.*

Remark 1. *Theorem 1 (ii) has been stated for the case* $m = p$*. Note that the case* $m > p$ *(i.e., there are more inputs than outputs) is irrelevant for the stabilization from the strict algebraic point of view since the* $(m - p)$ *extra inputs would be redundant. In the case that* $m \le p$*, Theorem 1 (ii) might be directly generalized to a subsystem´s decomposition philosophy if a number* $q \le m$ *of subsystems of inputs and outputs* $\left(u_1^T, u_2^T, \cdots, u_q^T\right)^T$ *and* $\left(y_1^T, y_2^T, \cdots, y_q^T\right)^T$ *with* $u_i \in \mathbf{R}^{m_i}$*,* $y_i \in \mathbf{R}^{p_i}$*;* $\forall i \in \overline{n}$ *with* $p = \sum_{i=1}^{q} p_i$ *and* $m = \sum_{i=1}^{q} m_i$*.*

Remark 2. *Theorem 1 can be easily generalized to cases when some dynamics transmission links between state, input or output components (or subsystems, in general) can be suppressed by manipulation. In more general cases, it is possible to extend Theorem 1 to combinations of the subsequent situations with the matrix decompositions having the same sense (in the various modified contexts) as that of Theorem 1 (ii):*

- **Case 1**. Suppression of some transmission links between the coupled open-loop dynamics by examining the decompositions: $A_c = A_{cd} + A_{cod}$, $A_d = A_{dd} + A_{dod}$, and $\widetilde{A} = A_c - A_d = (A_{cd} - A_{dd}) + (A_{cod} - A_{dod})$.

(*a*) If there is at least one non-diagonal zero entry in at least one of its rows in the off-diagonal controller error matrix $\widetilde{A}_{od} = A_{cod} - A_{dod}$ which is not a corresponding zero in A_{cod}; and (*b*) if there is at least one non-diagonal zero entry in at least one of its rows in the off-diagonal controller error matrix $\widetilde{K}_{od} = K_{cod} - K_{dod}$, then the closed-loop system is stabilizable under a partial decentralized control even if some links of the dynamics between crossed components are cut if Theorem 1 (ii) holds. If only Condition a is addressed, then the system is stabilizable by centralized control when cutting certain transmission links between coupled dynamics in the open-loop system. This idea can be extended to total decentralized control for a purely diagonal open-loop system´s dynamics under full zeroing of the off-diagonal corresponding error dynamics. It can be also generalized to the "ad hoc" decompositions between subsystems. Other cases with similar interpretations in the new contexts are:

- **Case 2**. Suppression of some crossed entries in the open-loop control matrix by examining the decompositions: $B_c = B_{cd} + B_{cod}$, $B_d = B_{dd} + B_{dod}$, and $\widetilde{B} = B_c - B_d = (B_{cd} - B_{dd}) + (B_{cod} - B_{dod})$.
- **Case 3**. Suppression of some crossed entries in the open-loop output matrix by examining the decompositions: $C_c = C_{cd} + C_{cod}$, $C_d = C_{dd} + C_{dod}$, and $\widetilde{C} = C_c - C_d = (C_{cd} - C_{dd}) + (C_{cod} - D_{dod})$.
- **Case 4**. Suppression of some crossed entries in the open-loop interconnection matrix by examining the decompositions: $D_c = D_{cd} + D_{cod}$, $D_d = D_{dd} + D_{dod}$, and $\widetilde{D} = D_c - D_d = (D_{cd} - D_{dd}) + (D_{cod} - D_{dod})$.
- **Case 5**. Any combinations of Cases 1–4.

Problem 1. *Find a stabilizing decentralized family of control gains by assuming that* $A_d = A_c$*, such that* $\widetilde{A} = A_c - A_d = 0$ *and Assumption 1 holds with* K_c *being a stabilizing centralized controller gain.*

The following more general result for the eventual case $\widetilde{A} \neq 0$ (that is $A_c \neq A_d$ eventually), follows from Theorem 1, Theorem A1 and Theorem A2 and Lemmas B1, B2 and B3:

Theorem 2. *Define the following error matrices between the centralized and decentralized system parameterizations:*

$$\widetilde{A} = A_c - A_d\,;\ \widetilde{B} = B_c - B_d;\ \widetilde{C} = C_c - C_d;\ \widetilde{D} = D_c - D_d;\ \widetilde{K} = K_c - K_d \tag{13}$$

such that $\|\widetilde{A}\| \leq \widetilde{\sigma}_A\varepsilon$, $\|\widetilde{B}\| \leq \widetilde{\sigma}_B\varepsilon$, $\|\widetilde{C}\| \leq \widetilde{\sigma}_c\varepsilon$, $\|\widetilde{D}\| \leq \widetilde{\sigma}_D\varepsilon$ *and* $\|\widetilde{K}\| \leq \widetilde{\sigma}_K\varepsilon$ *for some* $\varepsilon \in \mathbf{R}_{0+}$ *and given* $\widetilde{\sigma}_A, \widetilde{\sigma}_B, \widetilde{\sigma}_C, \widetilde{\sigma}_D \in \mathbf{R}_+$. *Assume that:*

(1) *Assumption 1 holds;*
(2) $K_c \in \mathbf{R}^{m\times p}$ *is a centralized linear output-feedback stabilizing controller gain such that the resulting closed-loop system matrix* $A_{cc} \in \mathbf{R}^{n\times n}$ *has a stability abscissa* $(-\rho_{cc}) < 0$ *and such that* $\|K_cD_c\|_2 < 1$ *(so that* $(I_m - K_cD_c)$ *is non-singular);*
(3) $A_c = A_d = A$; *and*
(4) *Define* $\overline{\varepsilon}^* = \min\left(1\,,\,\overline{\varepsilon},\,\overline{\varepsilon}_1\,,\,\overline{\varepsilon}_2\right)$, *where:*

$$\overline{\varepsilon} = \frac{\sqrt{\left(\widetilde{\sigma}_D\|K_c\| + \widetilde{\sigma}_K\|D_c\|\right)^2 + 4\widetilde{\sigma}_D\widetilde{\sigma}_K/\|(I_m - K_cD_c)^{-1}\|} - \left(\widetilde{\sigma}_D\|K_c\| + \widetilde{\sigma}_K\|D_c\|\right)}{2\widetilde{\sigma}_D\widetilde{\sigma}_K},$$

$$\overline{\varepsilon}_1 = \frac{\|(I_m - K_cD_c)^{-1}\|}{2[\|K_c\|\widetilde{\sigma}_D + \varepsilon\widetilde{\sigma}_D\widetilde{\sigma}_K + \|D_c\|\widetilde{\sigma}_K]};\ \overline{\varepsilon}_2 = 1\Big/\left(\widetilde{a}_{dc}\sup_{\omega\in\mathbf{R}_{0+}}\|(i\omega I_n - A_{cc})^{-1}\|_2\right),$$

where

$$\widetilde{a}_{dc} = (1 - \|K_cD_c\|)^{-1}\times[\widetilde{\sigma}_B\|K_cC_c\| + (\|B_c\| + \|K_cC_c\|)(\|K_c\|\widetilde{\sigma}_C + \|C_c\|\widetilde{\sigma}_K + C\,[\|K_c\|\widetilde{\sigma}_D + \|D_c\|\widetilde{\sigma}_K])],$$

where the non-negative real constant C is given in Equation (A17). Then, the following properties hold: ***(i)*** *If* $\widetilde{\sigma}_A = 0$ *(that is,* $A_c = A_d$*), then* A_{dc} *is stable and* $\varepsilon \in [0\,,\,\overline{\varepsilon}^*]$. ***(ii)*** *If* $\|\widetilde{A}\|_2 \leq \widetilde{\sigma}_A\varepsilon$, *then* A_{dc} *is stable and* $\varepsilon \in [0\,,\,\overline{\varepsilon}'^*]$ *where* $\overline{\varepsilon}'^* = \min\left(1\,,\,\overline{\varepsilon},\,\overline{\varepsilon}_1\,,\,\overline{\varepsilon}'_2\right)$ *and* $\overline{\varepsilon}'_2 = 1\Big/\left((\widetilde{a}_{dc} + \widetilde{\sigma}_A)\sup_{\omega\in\mathbf{R}_{0+}}\|(i\omega I_n - A_{cc})^{-1}\|_2\right)$.
(iii) *If* $\widetilde{A}(t)$ *is piecewise continuous and bounded, then Property (ii) holds by replacing* $\|\widetilde{A}\|_2 \leq \widetilde{\sigma}_A\varepsilon$ *by* $\sup_{0\leq t<\infty}\|\widetilde{A}(t)\|_2 \leq \widetilde{\sigma}_A\varepsilon$.

Remark 3. *Some quantified results are given in Lemmas B.2 and B.3 to modify* $\overline{\varepsilon}_2$*(and hence* $\overline{\varepsilon}'_2$*) in Theorem 2 by considering the second power of* ε *in the calculations of the disturbed parameterization guaranteeing the closed-loop stability in the decentralized case.*

Remark 4. *If the corresponding parametrical error matrices of Equation (13) have some zero off-diagonal entries (or off-diagonal block matrices in the more general case that the system is described by coupled subsystems), then we have at least a partial closed-loop stabilization under decentralized control or, eventually, cut coupled dynamic links to the light of the various Cases 1–5 described after Remark 2 such that closed-loop stability is preserved.*

Remark 5. *Theorem 2 also applies to the case of state-feedback control by replacing the output matrices* $C_c\,,\,C_d \to I_n$ *and fixing* $\widetilde{\sigma}_C = 0$.

Remark 6. *Theorem 2 also applies directly to the cases where Equations (1)–(3) are a given nominal asymptotically stable closed-loop system configuration and Equations (4)–(6) are a perturbed one whose closed-loop asymptotic stability maintenance related to its nominal counterpart is a suited objective and which is not necessarily of partial of compete decentralized type.*

4. Simulation Examples

This Section contains some numerical simulation examples to illustrate the theoretical results introduced in Section 3.

Example 1. *Consider the interconnected linear system with less inputs than outputs given by, [19]:*

$$\begin{aligned}\dot{x}_1(t) &= A_{11}x_1(t) + A_{12}x_2(t) + B_1u_1(t)\\ \dot{x}_2(t) &= A_{21}x_1(t) + A_{22}x_2(t) + B_2u_2(t)\\ y_1(t) &= C_1x_1(t)\\ y_2(t) &= C_2x_2(t)\end{aligned}$$

with $x_1(t)^T = \begin{bmatrix} x_{11}(t) & x_{12}(t) \end{bmatrix}$, $x_2(t)^T = \begin{bmatrix} x_{21}(t) & x_{22}(t) \end{bmatrix}$ *and matrices defined by:*

$$A_{11} = \begin{bmatrix} 1 & 0 \\ 0 & 2 \end{bmatrix}, A_{12} = A_{21} = \begin{bmatrix} -1 & 0 \\ 0 & -1 \end{bmatrix}, A_{22} = \begin{bmatrix} -1 & 0 \\ 0 & -2 \end{bmatrix}, B_1 = \begin{bmatrix} 1 \\ 2 \end{bmatrix}, B_2 = \begin{bmatrix} 2 \\ 1 \end{bmatrix}$$

$$C_1 = C_2 = I_2$$

This system can be cast into the form of Equations (1) and (2) by composing the matrices:

$$A_c = \begin{bmatrix} 1 & 0 & -1 & 0 \\ 0 & 2 & 0 & -1 \\ -1 & 0 & -1 & 0 \\ 0 & -1 & 0 & -2 \end{bmatrix}, B_c = \begin{bmatrix} 1 & 0 \\ 2 & 0 \\ 0 & 2 \\ 0 & 1 \end{bmatrix}, C_c = I_4, D_c = 0$$

Note that matrix A_c is unstable with eigenvalues given by $\{2.36, .1.41, -1.41, -2.23\}$. A static feedback output controller of the form of Equation (3) can be designed for this system, which leads to the following gain:

$$K_c = \begin{bmatrix} -9.3 & 8.2 & -0.015 & 0.02 \\ 0.01 & -0.01 & 0.375 & 0.25 \end{bmatrix}$$

that places the closed-loop poles at $\{-0.84, -1.1, -2.8, -3.3\}$ and thus stabilizes the closed-loop system. The static feedback gain K_c corresponds to a centralized controller as it can be readily observed. The question that arises now is whether a decentralized controller defined by:

$$K_d = \begin{bmatrix} -9.3 & 8.2 & 0 & 0 \\ 0 & 0 & 0.375 & 0.25 \end{bmatrix}$$

is enough to stabilize the system or not. Note that K_d is a block-diagonal matrix with zero off-block-diagonal entries. Theorem 1 enables us to guarantee the asymptotic stability of the above system when the decentralized controller K_d is used. Therefore, $A_d = A_c$, $B_d = B_c$ and $C_d = C_c$ while the feedback gain K_d is restricted to the proposed particular structure. In this way, consider now $A_{cc} = A_c + B_cK_cC_c$, $A_{dc} = A_c + B_cK_dC_c$ and $\widetilde{A}_{dc} = B_c(K_c - K_d)C_c$. Condition 2 of Theorem 1 *(i)* yields:

$$0.055 = \|\widetilde{A}_{dc}\|_2 < \frac{1}{\underset{\omega \in \mathbf{R}_{0+}}{sup}\ \|(i\omega I_4 - A_{cc})^{-1}\|_2} = 0.07$$

Consequently, we can conclude from Theorem 1 that the closed-loop system controlled by the decentralized static output gain K_d is asymptotically stable. Thus, we have been able to easily analyze the stability of the decentralized case from the stability property of the centralized one. Figure 1 shows the trajectories of the closed-loop system when the gain K_d is deployed with initial conditions given by $x_1(t)^T = \begin{bmatrix} -3 & -4 \end{bmatrix}$, $x_2(t)^T = \begin{bmatrix} 5 & 6 \end{bmatrix}$. It can be observed in Figure 1 that all the states converge to zero as predicted by Theorem 1.

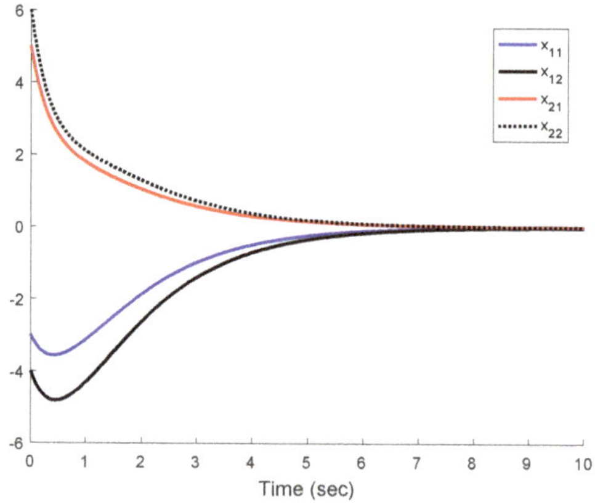

Figure 1. States evolution when the decentralized controller K_d is employed.

Example 2. *Consider the linear system with the same number of inputs and outputs composed of two identical pendulums THAT are coupled by a spring and subject to two distinct inputs, as displayed in Figure 2, [19].*

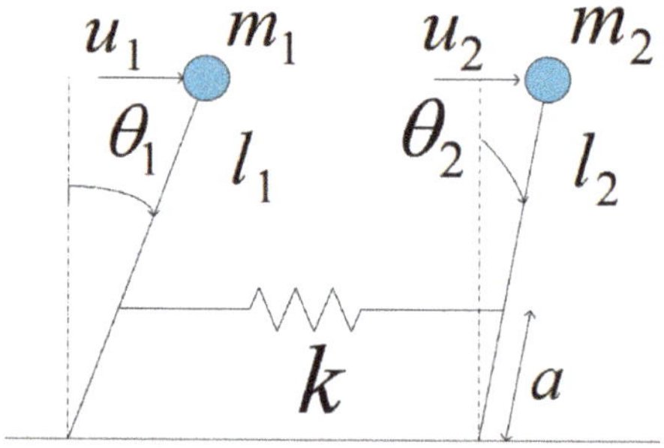

Figure 2. Two inverted pendulums coupled by a spring.

The mathematical model of such interconnected system is given by:

$$\begin{aligned} \dot{x}_1(t) &= A_{11}x_1(t) + A_{12}x_2(t) + B_1u_1(t) \\ \dot{x}_2(t) &= A_{21}x_1(t) + A_{22}x_2(t) + B_2u_2(t) \\ y_1(t) &= C_1x_1(t) \\ y_2(t) &= C_2x_2(t) \end{aligned}$$

with $x_1(t)^T = \begin{bmatrix} \theta_1 & \dot{\theta}_1 \end{bmatrix}$, $x_2(t)^T = \begin{bmatrix} \theta_2 & \dot{\theta}_2 \end{bmatrix}$ and matrices defined by:

$$A_{11} = A_{22} = \begin{bmatrix} 0 & 1 \\ \frac{g}{l} - \frac{ka^2}{ml^2} & -\mu \end{bmatrix}, A_{12} = A_{21} = \begin{bmatrix} 0 & 0 \\ \frac{ka^2}{ml^2} & 0 \end{bmatrix}, B_1 = B_2 = \begin{bmatrix} 0 \\ \frac{1}{ml^2} \end{bmatrix}$$

$$C_1 = C_2 = \begin{bmatrix} 1 & 0 \end{bmatrix}$$

where g represents the gravity, μ accounts for the friction, $m = m_1 = m_2$ are the masses of both pendulums, k is the spring constant and the meanings of the geometrical parameters are shown in Figure 2. This linear model corresponds to the linearization of the pendulum nonlinear equations around the up-right position equilibrium point. The following values were used in simulation, [19]:

$$\frac{g}{l} = 1, \frac{1}{ml^2} = 1, \mu = 1, \frac{k}{m} = 2, \frac{a}{l} = 0.5$$

This system can be cast into the form of Equations (1) and (2) as:

$$A_c = \begin{bmatrix} 0 & 1 & 0 & 0 \\ 0.5 & -1 & 0.5 & 0 \\ 0 & 0 & 0 & 1 \\ 0.5 & 0 & 0.5 & -1 \end{bmatrix}, B_c = \begin{bmatrix} 0 & 0 \\ 1 & 0 \\ 0 & 0 \\ 0 & 1 \end{bmatrix}, C_c = \begin{bmatrix} 1 & 0 & 0 & 0 \\ 0 & 0 & 1 & 0 \end{bmatrix}, D_c = 0$$

A static output feedback controller can be designed for this system to achieve its asymptotic stability. In this way, the feedback gain

$$K_c = \begin{bmatrix} 2.92 & 0.65 \\ 0.64 & 2.80 \end{bmatrix}$$

places the closed-loop poles at $\{-0.5 \pm 1.5i, -0.5 \pm 1.4i\}$ with negative real parts. Now, we implement a decentralized controller with feedback gain given by:

$$K_d = \begin{bmatrix} 2.92 & 0 \\ 0 & 2.80 \end{bmatrix}$$

Theorem 2 is now used to analyze the stability of the closed-loop system when this controller is employed. This case is of practical importance and corresponds to the situation when the local controller has only available for control purposes the information regarding the local output, and not the output of the complete system. Thus, the centralized and decentralized systems are the same and only the static feedback gain changes. Theorem 2 conditions are applied with $\widetilde{\sigma}_A = \widetilde{\sigma}_B = \widetilde{\sigma}_C = \widetilde{\sigma}_D = 0$, $\widetilde{\sigma}_K = ||K_c - K_d||_2 = 0.65$ while the stability condition for this special case (see Appendix B) reads:

$$0.65 = 0.65 \times 1 \times 1 = ||K_c - K_d||_2 ||B_c||_2 ||C_c||_2 < 1$$

Accordingly, the closed-loop system attained with the decentralized controller is asymptotically stable and all the outputs will converge to zero asymptotically. Figure 3 displays the evolution of both angles from initial conditions $x_1(t)^T = \begin{bmatrix} 0.5 & -0.5 \end{bmatrix}$, $x_2(t)^T = \begin{bmatrix} 0.15 & 0.5 \end{bmatrix}$, where it can be observed that both pendulums are stabilized in the up-right position with the decentralized controller.

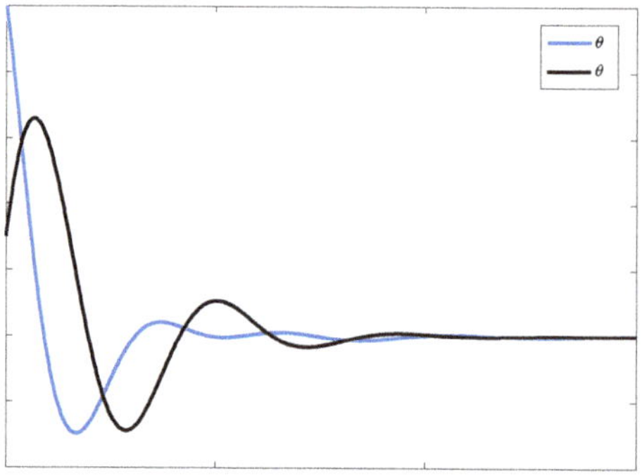

Figure 3. Evolution of the angles of both pendulums.

Example 3. *Consider the linear interconnected system given by:*

$$\begin{aligned} \dot{x}_c(t) &= A_c x_c(t) + B_c u_c(t) \\ y_c(t) &= C_c x_c(t) + D_c u_c(t) \end{aligned}$$

with matrices defined by:

$$A_c = \begin{bmatrix} -1 & 0.4 & 0.3 \\ 0.2 & -2 & 0.1 \\ -0.1 & 0.2 & -3 \end{bmatrix}, B_c = \begin{bmatrix} 1 & 0.1 & 0 \\ 0 & 2 & 0 \\ 0 & 0 & 2 \end{bmatrix}, C_c = \begin{bmatrix} 1 & 0.1 & 0 \\ 0 & 1 & 0.1 \\ 0 & 0 & 1 \end{bmatrix}, D_c = 0.1 I_3$$

This system is controlled by the static output feedback gain given by:

$$K_c = \begin{bmatrix} 0.19 & 0.05 & 0.04 \\ 0.05 & -0.01 & 0.03 \\ -0.02 & 0.02 & -0.46 \end{bmatrix}$$

which places the closed-loop poles at $\{-1.17, -2, -2.1\}$. The decentralized system is now given by:

$$A_d = A_c, B_d = \begin{bmatrix} 1 & 0 & 0 \\ 0 & 2 & 0 \\ 0 & 0 & 2 \end{bmatrix}, C_d = I_3, D_d = 0$$

The decentralized system corresponds to the case when some transmission links have been suppressed from the original open-loop coupled dynamics, as considered in Remark 2. The following decentralized gain iz employed to stabilize the decentralized system in Equations (4)–(6) parameterized by the above matrices:

$$K_d = \begin{bmatrix} 0.19 & 0 & 0 \\ 0 & -0.01 & 0 \\ 0 & 0 & -0.46 \end{bmatrix}$$

Theorem 2 is used to analyze the stability of the decentralized closed-loop system. To this end, we calculate:

$$\|A_c - A_d\|_2 = \widetilde{\sigma}_A = 0$$

$$0.1 = \|B_c - B_d\|_2 \leq \widetilde{\sigma}_B \varepsilon = 2.7 \times 0.07 = 0.19$$

$$0.1 = \|C_c - C_d\|_2 \leq \widetilde{\sigma}_C \varepsilon = 2.7 \times 0.07 = 0.19$$

$$0.1 = ||D_c - D_d||_2 \leq \widetilde{\sigma}_D \varepsilon = 2.7 \times 0.07 = 0.19$$

$$0.07 = ||K_c - K_d||_2 \leq \widetilde{\sigma}_K \varepsilon = 2.7 \times 0.07 = 0.19$$

With these values, we can compute $\overline{\varepsilon} = 0.31$, $\overline{\varepsilon}_1 = 0.24$, $\overline{\varepsilon}_2 = 0.071$ so that $0.07 = \varepsilon < \overline{\varepsilon}^* = min(1, \overline{\varepsilon}, \overline{\varepsilon}_1, \overline{\varepsilon}_2) = 0.071$. Since $||K_c D_c||_2 = 0.05 < 1$, we are in conditions of applying Theorem 2 *(i)* and we can conclude that the decentralized closed-loop system is asymptotically stable. In this way, the presented results allow establishing the stability of the decentralized system by a simple method based on the stability and design of the centralized system. Figure 4 shows the state variables evolution from the initial state $x(t)^T = \begin{bmatrix} 5 & -5 & 1 \end{bmatrix}$. As shown in Figure 4, the state variables converge to zero asymptotically, as concluded from Theorem 2.

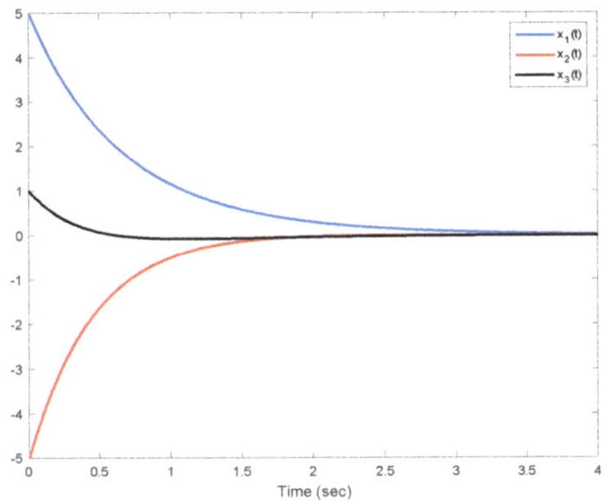

Figure 4. Evolution of the state space variables when the decentralized controller K_d is used.

5. Concluding Remarks

This paper is devoted to formulating sufficiency-type linear-output feedback decentralized closed-loop stabilization conditions if the continuous-time linear dynamic system can be stabilized under linear output-feedback centralized stabilization. The developed stability tests are conceptually simple to evaluate and they rely on the quantification in terms of worst-case norms of interconnection and open-loop system dynamics matrices and the corresponding control gains in the decentralized case compared to the centralized counterpart. The tolerances of the various parametrical matrix errors have been quantified by considering the first or second powers of a small parameter. Such a parameter is a design factor to characterize in the worst-case for the allowed tolerances to the perturbed parameterization norms. Simulated examples are discussed to illustrate the obtained results. The decentralized control design versus its decentralized control counterpart, under eventual output linear feedback, has been studied from the point of view of the amount of information that can be lost or omitted in terms of the total or partial knowledge of the coupled dynamics between subsystems necessary in the decentralized case to keep the closed-loop stability. A foreseen related future work relies on the application of the method to some applied control problems such as consensus protocols under decentralized control and continuous-discrete hybrid controller designs.

Author Contributions: Author M.D.l.S. contributed to the manuscript in the aspects of conceptualization; methodology; formal analysis; investigation; writing original draft preparation; writing review and editing; supervision and fund acquisition. Author A.I. contributed to the manuscript in the various aspects of preparation of the numerical experiments; software implementation; visualization and original draft preparation.

Funding: This research was co-funded by the Spanish Government and the European Fund of Regional Development FEDER, grant numbers DPI2015-64766-R and DPI2016-77271-R (MINECO/FEDER, UE) and the University of the Basque Country UPV /EHU, grant PGC 17/33.

Acknowledgments: The authors are grateful to the Spanish Government for grants DPI2015-64766-R and DPI2016-77271-R (MINECO/FEDER, UE), and to UPV /EHU for grant PGC 17/33. They also thank the referees for their useful comments.

Conflicts of Interest: The authors declare no conflict of interest.

Appendix A *Auxiliary Stability Results on Perturbed Matrices under Constant and Time-Varying Perturbations of Stability Matrices*

Theorem A1. *Assume that $A_0 \in \mathbf{R}^{n\times n}$ is a stability matrix with stability abscissa $(-\rho_c) < 0$. Then, $A = A_0 + \widetilde{A}$ is a stability matrix if any of the subsequent sufficiency-type conditions holds:*

(1) The H_∞-norm of $(sI_n - A_0)^{-1}\widetilde{A}$ satisfies $\|(sI_n - A_0)^{-1}\widetilde{A}\|_\infty < 1$,
(2) $\|\widetilde{A}\|_2 < 1/\sup_{\omega\in\mathbf{R}_{0+}} \|(i\omega I_n - A_0)^{-1}\|_2$.
Other alternative sufficiency-type conditions to Conditions 1 and 2 for the stability of A are:
(3) $\rho\left(A_0^{-1}\widetilde{A}\right) < 1$,
(4) $\|A_0^{-1}\widetilde{A}\|_2 < 1$,

$\|\widetilde{A}\|_2 < 1/\|A_0^{-1}\|_2$, that is, $\lambda_{\max}\left(\widetilde{A}^T\widetilde{A}\right) < \lambda_{\min}\left(A_0^T A_0\right)$, in the following particular cases:
(a) $A_0 \prec 0$ and $A \prec -A_0$; and
(b) $A_0 = \left(A_{0_{ij}}\right) \in M_E^{n\times n}$ and $\widetilde{A} = \left(\widetilde{A}_{ij}\right)$ fulfils $\widetilde{A}_{ij} \geq -A_{0_{ij}}$; $\forall i, j(\neq i) \in \overline{n}$.

Proof. Note that

$$\begin{aligned}\det(sI_n - A) &= \det\left(sI_n - A_0 - \widetilde{A}\right) = \det\left((sI_n - A_0)\left(I_n - (sI_n - A_0)^{-1}\widetilde{A}\right)\right)\\ &= \det(sI_n - A_0)\ \det\left(I_n - (sI_n - A_0)^{-1}\widetilde{A}\right); \forall s \in \mathbf{C}\end{aligned} \tag{A1}$$

and then $\det(sI_n - A) = \det(sI_n - A_0)\ \det\left(I_n - (sI_n - A_0)^{-1}\widetilde{A}\right) \neq 0$ Then, for all $s \notin sp(A)$, and also for all $s \in sp(A_0)$ if the H_∞-norm of $(sI_n - A_0)^{-1}\widetilde{A}$, which exists since A_0 is a stability matrix, satisfies $\|(sI_n - A_0)^{-1}\widetilde{A}\|_\infty < 1$, which is guaranteed if $\|\widetilde{A}\|_2 < 1/\sup_{\omega\in\mathbf{R}_{0+}} \|(i\omega I_n - A_0)^{-1}\|_2$. Then, A is a stability matrix if Conditions 1 or 2 holds. On the other hand, if A_0 and A are negative (implying that $A \prec -A_0$), or if they are both Metzler-stable (implying for all off-diagonal entries that $\widetilde{A}_{ij} \geq -A_{0_{ij}}$; $\forall i, j(\neq i) \in \overline{n}$), then their dominant abscissa (perhaps multiple) eigenvalue is real and negative since A_0 being a stability matrix is claimed to guarantee that A is stable. Since A_0 is a stability matrix, it is non-singular with eigenvalues with negative real parts. Then, by the continuity of the eigenvalues with respect to the matrix entries, $A = A_0\left(I_n + A_0^{-1}\widetilde{A}\right)$ is a stability matrix if $\rho\left(A_0^{-1}\widetilde{A}\right) \leq \|A_0^{-1}\widetilde{A}\|_2 \leq \|A_0^{-1}\|_2\|\widetilde{A}\|_2 < 1$ leading to the sufficiency of Conditions 3–5 for the stability of A if is A_0 stable. The last, sufficient condition comes directly by upper-bounding Condition 4 by norm product and it is equivalent to $\|\widetilde{A}\|_2 = \lambda_{\max}^{1/2}\left(\widetilde{A}^T\widetilde{A}\right) < 1/\|A_0^{-1}\|_2 = 1/\lambda_{\max}^{1/2}\left(A_0^{-1}A_0^{-T}\right) = \lambda_{\min}^{1/2}\left(A_0^T A_0\right)$. □

Theorem A2. *Assume that $A_0 \in \mathbf{R}^{n\times n}$ is a stability matrix and that $A(t) = A_0 + \widetilde{A}(t)$, where $\widetilde{A} : \mathbf{R}_{0+} \to \mathbf{R}^{n\times n}$ is piecewise-continuous and bounded. Then, $A : \mathbf{R}_{0+} \to \mathbf{R}^{n\times n}$ is stable if $\frac{k_0}{\rho_0}\sup_{0\leq\tau\leq t}\|\widetilde{A}(\tau)\| < 1$; $\forall t \in \mathbf{R}_{0+}$, where $(-\rho_0) < 0$ is the stability abscissa of A_0 and $k_0 = k_0(\|A_0\|) \geq 1$ is a real constant satisfying $\|e^{A_0 t}\| \leq k_0 e^{-\rho_0 t}$; $\forall t \in \mathbf{R}_{0+}$.*

Proof. Consider the linear time-varying system:

$$\dot{x}(t) = \left(A_0 + \widetilde{A}(t)\right)x(t),\ x(0) = x_0\ ;\ \forall t \in \mathbf{R}_{0+} \tag{A2}$$

where $\widetilde{A} : \boldsymbol{R}_{0+} \to \boldsymbol{R}^{n \times n}$ is piecewise-continuous and bounded [1]. Such a system is globally asymptotically stable if and only if $A : \boldsymbol{R}_{0+} \to \boldsymbol{R}^{n \times n}$ is a stability matrix. The state-trajectory solution of Equation (A2) satisfies:

$$\begin{aligned} \|x(t)\| &\leq \|e^{A_0 t}\| \|x_0\| + \int_0^t \left\|e^{A_0(t-\tau)}\right\| \|\widetilde{A}(\tau)\| \|x(\tau)\| d\tau \leq k_0 e^{-\rho_0 t}\|x_0\| + \frac{k_0}{\rho_0} \sup_{0\leq\tau\leq t} \|x(\tau)\| \\ &= K_0 e^{-\rho_0 t}\|x_0\| + \frac{k_0}{\rho_0} \sup_{0\leq\tau\leq t} \left(\|\widetilde{A}(\tau)\|\right) \|x(t\prime)\|; \forall t \in \boldsymbol{R}_{0+} \end{aligned} \quad \text{(A3)}$$

Let $t\prime = t\prime(t)$ be defined for each $t \in \boldsymbol{R}_{0+}$ as $t\prime = \left\{z = \max_{0\leq\tau\leq t} \tau \ : \ \|x(z)\| \geq \|x(t)\|\right\}$. Then,

$$\|x(t)\| \leq \sup_{0\leq\tau\leq t} \|x(\tau)\| = \sup_{0\leq\tau\leq t\prime} \|x(\tau)\| = \|x(t\prime)\| \leq k_0{}^{-\rho_0 t\prime}\|x_0\| + \frac{k_0}{\rho_0} \sup_{0\leq\tau\leq t} \|\widetilde{A}(\tau)\| \|x(t\prime)\|; \forall t \in \boldsymbol{R}_{0+} \quad \text{(A4)}$$

Since $1 > \frac{k_0}{\rho_0} \sup_{0\leq\tau\leq t} \|\widetilde{A}(\tau)\|; \forall t \in \boldsymbol{R}_{0+}$, one gets:

$$\begin{aligned} \| x(t)\| &\leq \| x(t\prime)\| = \sup_{0\leq\tau\leq t\prime} \|x(\tau)\| \leq \left(1 - \frac{k_0}{\rho_0} \sup_{0\leq\tau<\infty} \|\widetilde{A}(\tau)\|\right)^{-1} k_0 e^{-\rho_0 t\prime}\|x_0\| \\ &\leq M = k_0\left(1 - \frac{k_0}{\rho_0} \sup_{0\leq\tau<\infty} \|\widetilde{A}(\tau)\|\right)^{-1} \|x_0\| \ ; \ \forall t \in \boldsymbol{R}_{0+} \end{aligned} \quad \text{(A5)}$$

Therefore, $x(t)$ is bounded for any $t \in \boldsymbol{R}_{0+}$ if x_0 is finite. Now, assume the following cases.

Case a: For any $T_s \in \boldsymbol{R}_+$, the sequence $\left\{\sup_{nT_s\leq t<(n+m_n)T_s} \|x(t)\|\right\}_{n=k}^{\infty}$ is strictly decreasing for some finite positive integer $k = k(T_s)$ and some positive sequence $\{m_n\}$ of bounded integer numbers which satisfies $m_{n+1} > m_n - 1$ for $n \geq 0$. As a result, $\left\{\sup_{nT_s\leq t<(n+m_n)T_s} \|x(t)\|\right\}_{n=k}^{\infty} \to 0$ as $n \to \infty$ for any given $T_s > 0$. Then, one gets from Equation (A3) that $\|x(t)\| \to 0$, as $t \to \infty$ since $\|x(nT_s + t)\| \leq k_0\left(1 + \frac{1}{\rho_0} \sup_{nT_s\leq\tau<t} \|x(\tau)\|\right); \forall t \in (nT_s\,,\, (n+1)T_s)$. The result is proved for this case.

Case b: For some $T_s \in \boldsymbol{R}_+$, a sequence $\left\{\sup_{nT_s\leq t<(n+m_n)T_s} \|x(t)\|\right\}_{n=0}^{\infty}$ can be built, with $\{m_n\}_{n=0}^{\infty} \to \infty$ as $m \to \infty$ satisfying $\sup_{(n+m_n)T_s\leq t<(n+m_n)T_s} \|x(t)\| = \sup_{nT_s\leq t<(n+m_n+m_{n+m_n})T_s} \|x(t)\| \geq \sup_{nT_s\leq t<(n+m_n)T_s} \|x(t)\| \ ; n \geq 0$ (note that the above inequality cannot be strict as $n \to \infty$ since it has already been proven that $\|x(t)\| < +\infty; \forall t \in \boldsymbol{R}_{0+}$). However, then one gets from Equation (A3) for some $t_{n+1} \in [n + m_n\,,\, n + m_n + m_{n+m_n})$ since $1 > \frac{k_0}{\rho_0} \sup_{0\leq\tau\leq t} \|\widetilde{A}(\tau)\|; \forall t \in \boldsymbol{R}_{0+}$:

$$\begin{aligned} &\sup_{nT_s\leq t<(n+m_n)T_s} \|x(t)\| \leq \sup_{nT_s\leq t<(n+m_n+m_{n+m_n})T_s} \|x(t)\| \leq K_0 e^{-\rho_0(t_{n+1}-nT_s)}\|x(nT_s)\| \\ &+ \frac{K_0}{\rho_0} \sup_{nT_s\leq t<(n+m_n+m_{n+m_n})T_s} \left(\|\widetilde{A}(\tau)\|\right) \sup_{nT_s\leq t<(n+m_n+m_{n+m_n})T} \|x(t)\| \\ &< K_0 e^{-\rho_0(t_{n+1}-nT_s)}\|x(nT_s)\| + \sup_{nT_s\leq t<(n+m_n+m_{n+m_n})T} \|x(t)\| \end{aligned} \quad \text{(A6)}$$

If $n \to \infty$, $m_n \to \infty$, then $(t_{n+1} - m_{n+m_n}) \to \infty$, thus the following contradiction arises:

$$0 = \limsup_{n\to\infty} \left(\sup_{nT_s\leq t<(n+m_n+m_{n+m_n})T_s} \|x(t)\| - \sup_{nT_s\leq t<(n+m_n+m_{n+m_n})T_s} \|x(t)\|\right) < 0.$$

Thus, Case b is not possible and the whole result follows from Case a. □

Note that the stability abscissa of A_0, that is, $(-\rho_0) < 0$ is not smaller than the dominant eigenvalue abscissa.

Appendix B *Calculations for solving Problem 1*

Assume that the matrix $(I_m - K_c D_c)$ is non-singular with $\|K_c D_c\|_2 < 1$ and $A = A_c = A_d$. Then, one gets from Equations (10) and (13) that:

$$\begin{aligned} & A_d + B_d (I_m - K_d D_d)^{-1} K_d C_d \\ & = A + \left(B_c - \widetilde{B}\right)\left(I_m - K_c D_c - \widetilde{\Delta}_0\right)^{-1}\left(K_c D_c - \widetilde{\Delta}_1\right) \\ & = A + \left(B_c - \widetilde{B}\right)\left[(I_m - K_c D_c)\left(I_m - (I_m - K_c D_c)^{-1}\widetilde{\Delta}_0\right)\right]^{-1}\left(K_c C_c - \widetilde{\Delta}_1\right) \\ & = A + \left(B_c - \widetilde{B}\right)\left(I_m - (I_m - K_c D_c)^{-1}\widetilde{\Delta}_0\right)^{-1}(I_m - K_c D_c)^{-1}\left(K_c C_c - \widetilde{\Delta}_1\right) \\ & = A + \left(B_c - \widetilde{B}\right)\left(I_m + \widetilde{\Delta}_2\right)(I_m - K_c D_c)^{-1}\left(K_c C_c - \widetilde{\Delta}_1\right) \end{aligned} \tag{A7}$$

provided that $\widetilde{D}$ and $\widetilde{K}$ are such that $(I_m - K_d D_d)^{-1} = \left(I_m - K_c D_c - \widetilde{\Delta}_0\right)^{-1}$ exists (note that this always holds if $\widetilde{D} = 0_{p\times m}$ and $\widetilde{K} = 0_{m\times p}$ from Assumption 1), where:

$$\widetilde{\Delta}_0 = K_d D_d - K_c D_c = \widetilde{K}\left(\widetilde{D} - D_c\right) - K_c\widetilde{D} = \left(\widetilde{K} - K_c\right)\widetilde{D} - \widetilde{K}D_c \tag{A8}$$

$$\widetilde{\Delta}_1 = \widetilde{K}\left(C_c - \widetilde{C}\right) + K_c\widetilde{C} = \left(K_c - \widetilde{K}\right)\widetilde{C} + \widetilde{K}C_c \tag{A9}$$

$$\widetilde{\Delta}_2 = \left[I_m - (I_m - K_c D_c)^{-1}\widetilde{\Delta}_0\right]^{-1} - I_m \tag{A10}$$

and note that

$$\|\widetilde{\Delta}_0\| \le \varepsilon\left[\|K_c\|\widetilde{\sigma}_D + \varepsilon\widetilde{\sigma}_D\widetilde{\sigma}_K + \|D_c\|\widetilde{\sigma}_K\right] \tag{A11}$$

$$\|\widetilde{\Delta}_1\| \le \varepsilon\left[\|K_c\|\widetilde{\sigma}_C + \varepsilon\widetilde{\sigma}_C\widetilde{\sigma}_K + \|C_c\|\widetilde{\sigma}_K\right] \tag{A12}$$

and one gets from Banach´s Perturbation Lemma [7] that

$$\|\widetilde{\Delta}_2\| \le 1 + \frac{1}{1 - \|(I_m - K_c D_c)^{-1}\widetilde{\Delta}_0\|} \le 1 + \frac{1}{1 - \|(I_m - K_c D_c)^{-1}\|\,\|\widetilde{\Delta}_0\|} \tag{A13}$$

provided that $\|\widetilde{\Delta}_0\| < \widetilde{\delta}_0 = 1/\|(I_m - K_c D_c)^{-1}\|$. Equivalently, if

$$q(\varepsilon) = \widetilde{\sigma}_D\widetilde{\sigma}_k\varepsilon^2 + (\widetilde{\sigma}_D\|K_c\| + \widetilde{\sigma}_K\|D_c\|)\varepsilon - 1/\|(I_m - K_c D_c)^{-1}\| < 0 \tag{A14}$$

since $q(\varepsilon)$ is a convex parabola with zeros $\varepsilon_1 < 0$ and $\overline{\varepsilon} = \varepsilon_2 > 0$, Equation (A14) holds, guaranteeing that $\|\widetilde{\Delta}_0\| < \widetilde{\delta}_0$, if $\varepsilon \in [0, \overline{\varepsilon})$, where:

$$\overline{\varepsilon} = \frac{\sqrt{(\widetilde{\sigma}_D\|K_c\| + \widetilde{\sigma}_K\|D_c\|)^2 + 4\widetilde{\sigma}_D\widetilde{\sigma}_K/\|(I_m - K_c D_c)^{-1}\|} - (\widetilde{\sigma}_D\|K_c\| + \widetilde{\sigma}_K\|D_c\|)}{2\widetilde{\sigma}_D\widetilde{\sigma}_K} \tag{A15}$$

Before continuing with the calculations, we give the following auxiliary result:

Lemma B1. *If $(I_m - K_c D_c)$ is non-singular with $\|K_c D_c\|_2 < 1$ and $\|\widetilde{\Delta}_0\| < \widetilde{\delta}_0 = 1/\|(I_m - K_c D_c)^{-1}\|$, equivalently if $\varepsilon \in [0, \overline{\varepsilon})$, with $\overline{\varepsilon}$ defined in Equation (A15), then $\|\widetilde{\Delta}_2\| \le C\|\widetilde{\Delta}_0\| < 1$ with a norm-dependent real constant $C \ge \frac{1}{2\widetilde{\delta}_0}$ if $\varepsilon \in [0, \overline{\varepsilon}_1)$ with*

$$\overline{\varepsilon}_1 = \frac{\|(I_m - K_c D_c)^{-1}\|}{2\left[\|K_c\|\widetilde{\sigma}_D + \varepsilon\widetilde{\sigma}_D\widetilde{\sigma}_K + \|D_c\|\widetilde{\sigma}_K\right]} \tag{A16}$$

Proof. One gets from Equation (A10) and Banach´s Perturbation Lemma [7] that, if $\|\widetilde{\Delta}_2\| \leq C\|\widetilde{\Delta}_0\|$ for some $C \in \boldsymbol{R}_+$, then:

$$\frac{1}{1-C\|\widetilde{\Delta}_0\|} \geq \|\left(\widetilde{\Delta}_2+I_m\right)^{-1}\| = \|I_m-(I_m-K_cD_c)^{-1}\widetilde{\Delta}_0\| \geq 1-\|(I_m-K_cD_c)^{-1}\|\,\|\widetilde{\Delta}_0\|$$

provided that $C < 1/\|\widetilde{\Delta}_0\|$. One gets that the above inequality holds if $1/\|\widetilde{\Delta}_0\| > C \geq \frac{1}{2\delta_0} \geq \frac{\|(I_m-K_cD_c)^{-1}\|}{1+\|\widetilde{\Delta}_0\|\,\|(I_m-K_cD_c)^{-1}\|}$ and, one gets from Equation (A11) that

$$\|\widetilde{\Delta}_2\| \leq \varepsilon\, C\left[\|K_c\|\widetilde{\sigma}_D+\varepsilon\widetilde{\sigma}_D\widetilde{\sigma}_K+\|D_c\|\widetilde{\sigma}_K\right] < 1 \tag{A17}$$

if $\overline{\varepsilon} < \overline{\varepsilon}_1$. □

Now, rewrite the system matrices of closed-loop dynamics of Equations (7) and (10), equivalently Equation (A7), with $A = A_c = A_d$ and its incremental value as follows:

$$A_{cc} = A + B_c(I_m-K_cD_c)^{-1}K_cC_c \tag{A18}$$

$$A_{dc} = A + B_d(I_m-K_dD_d)^{-1}K_dC_d \tag{A19}$$

$$\begin{aligned}
\widetilde{A}_{dc} &= A_{cc}-A_{dc} = B_c(I_m-K_cD_c)^{-1}K_cC_c - B_d(I_m-K_dD_d)^{-1}K_dC_d \\
&= B_c(I_m-K_cD_c)^{-1}K_cC_c - \left(B_c-\widetilde{B}\right)\left(I_m+\widetilde{\Delta}_2\right)(I_m-K_cD_c)^{-1}\left(K_cC_c-\widetilde{\Delta}_1\right) \\
&= B_c(I_m-K_cD_c)^{-1}K_cC_c - B_c\left(I_m+\widetilde{\Delta}_2\right)(I_m-K_cD_c)^{-1}\left(K_cC_c-\widetilde{\Delta}_1\right) \\
&+\widetilde{B}\left(I_m+\widetilde{\Delta}_2\right)(I_m-K_cD_c)^{-1}\left(K_cC_c-\widetilde{\Delta}_1\right) = B_c(I_m-K_cD_c)^{-1}\widetilde{\Delta}_1 \\
&-B_c\widetilde{\Delta}_2(I_m-K_cD_c)^{-1}K_cC_c + B_c\widetilde{\Delta}_2(I_m-K_cD_c)^{-1}\widetilde{\Delta}_1 \\
&+\widetilde{B}(I_m-K_cD_c)^{-1}K_cC_c - \widetilde{B}(I_m-K_cD_c)^{-1}\widetilde{\Delta}_1 \\
&+\widetilde{B}\widetilde{\Delta}_2(I_m-K_cD_c)^{-1}K_cC_c - \widetilde{B}\widetilde{\Delta}_2(I_m-K_cD_c)^{-1}\widetilde{\Delta}_1
\end{aligned} \tag{A20}$$

Now, the following technical result follows directly from Equations (A20), (A11), (A12) and (A17), the norm upper-bounding values of the control, output interconnection and controller matrix errors and Lemma B1:

Lemma B2. *The following properties hold for any $\varepsilon \in [0, \hat{\varepsilon})$ with $\hat{\varepsilon} = \min(\overline{\varepsilon}, \overline{\varepsilon}_1)$ calculated from Equations (A15) and (A16):*

(i)

$$\begin{aligned}
&\|\widetilde{A}_{dc}\| \leq \|B_c\|(1-\|K_cD_c\|)^{-1}\left[\|\widetilde{\Delta}_1\|\left(1+\|\widetilde{\Delta}_2\|\right)+\|K_cC_c\|\|\widetilde{\Delta}_2\|\right] \\
&+\varepsilon\,(1-\|K_cD_c\|)^{-1}\widetilde{\sigma}_B\left(1+\|\widetilde{\Delta}_2\|\right)\left(\|K_cC_c\|+\|\widetilde{\Delta}_1\|\right) \leq \varepsilon\|B_c\|(1-\|K_cD_c\|)^{-1} \\
&\times[(\|K_c\|\widetilde{\sigma}_C+\varepsilon\widetilde{\sigma}_C\widetilde{\sigma}_K+\|C_c\|\widetilde{\sigma}_K)(1+\|K_cC_c\|+\varepsilon C\left[\|K_c\|\widetilde{\sigma}_D+\varepsilon\widetilde{\sigma}_D\widetilde{\sigma}_K+\|D_c\|\widetilde{\sigma}_K\right])C\left[\|K_c\|\widetilde{\sigma}_D+\|D_c\|\widetilde{\sigma}_K+\varepsilon\widetilde{\sigma}_D\widetilde{\sigma}_K\right]] \\
&+\varepsilon\,(1-\|K_cD_c\|)^{-1}\widetilde{\sigma}_B(1+\varepsilon\, C\left[\|K_c\|\widetilde{\sigma}_D+\varepsilon\widetilde{\sigma}_D\widetilde{\sigma}_K+\|D_c\|\widetilde{\sigma}_K\right])(\|K_cC_c\|+\varepsilon\left[\|K_c\|\widetilde{\sigma}_C+\varepsilon\widetilde{\sigma}_C\widetilde{\sigma}_K+\|C_c\|\widetilde{\sigma}_K\right])
\end{aligned} \tag{A21}$$

(ii) *If, furthermore, $\varepsilon \leq 1$, then $\varepsilon^r \leq \varepsilon$ for any real $r \geq 1$ so that one gets from Equation (A21) by taking the upper-bound ε^2 for ε^3 that*

$$\begin{aligned}
&\|\widetilde{A}_{dc}\| \leq \varepsilon\|B_c\|(1-\|K_cD_c\|)^{-1}\left(\|\widetilde{\Delta}_{10}\|+\|K_cC_c\|\|\widetilde{\Delta}_{20}\|\right)+\varepsilon^2\|\widetilde{\Delta}_{20}\|\,\|B_c\|(1-\|K_cD_c\|)^{-1} \\
&+\varepsilon\,(1-\|K_cD_c\|)^{-1}\widetilde{\sigma}_B\|K_cC_c\|+\varepsilon^2\,(1-\|K_cD_c\|)^{-1}\widetilde{\sigma}_B\left(\|\widetilde{\Delta}_{10}\|+\|K_cC_c\|\|\widetilde{\Delta}_{20}\|\right) \\
&+\varepsilon^3\,(1-\|K_cD_c\|)^{-1}\widetilde{\sigma}_B\|\widetilde{\Delta}_{20}\|\|\widetilde{\Delta}_{10}\| \leq u\varepsilon+v\varepsilon^2 \leq (u+v)\varepsilon
\end{aligned} \tag{A22}$$

where $\widetilde{\Delta}_i = \varepsilon\widetilde{\Delta}_{i0}$ *for* $i = 1, 2$ *and it has been used that* $\varepsilon^3 \leq \varepsilon^2$, *with*

$$\begin{aligned} u &= \|B_c\|(1-\|K_cD_c\|)^{-1}\left(\|\widetilde{\Delta}_{10}\| + \|K_cC_c\|\|\widetilde{\Delta}_{20}\|\right) + (1-\|K_cD_c\|)^{-1}\widetilde{\sigma}_B\|K_cC_c\| \\ v &= (1-\|K_cD_c\|)^{-1}\widetilde{\sigma}_B\|\widetilde{\Delta}_{20}\|\|\widetilde{\Delta}_{10}\| + \|\widetilde{\Delta}_{20}\|\,\|B_c\|(1-\|K_cD_c\|)^{-1} + (1-\|K_cD_c\|)^{-1}\widetilde{\sigma}_B\left(\|\widetilde{\Delta}_{10}\| + \|K_cC_c\|\|\widetilde{\Delta}_{20}\|\right) \end{aligned} \tag{A23}$$

(iii) *If the upper-bound* ε *is used for* ε^2 *and* ε^3, *one gets that*

$$\begin{aligned} &\|\widetilde{A}_{dc}\| \leq \varepsilon\|B_c\|(1-\|K_cD_c\|)^{-1} \\ &\times[(\|K_c\|\widetilde{\sigma}_C + \widetilde{\sigma}_C\widetilde{\sigma}_K + \|C_c\|\widetilde{\sigma}_K)(1 + C\,[\|K_c\|\widetilde{\sigma}_D + \widetilde{\sigma}_D\widetilde{\sigma}_K + \|D_c\|\widetilde{\sigma}_K]) + \|K_cC_c\|C\,[\|K_c\|\widetilde{\sigma}_D + \widetilde{\sigma}_D\widetilde{\sigma}_K + \|D_c\|\widetilde{\sigma}_K]] \\ &+\varepsilon\,(1-\|K_cD_c\|)^{-1}\widetilde{\sigma}_B(1 + C\,[\|K_c\|\widetilde{\sigma}_D + \widetilde{\sigma}_D\widetilde{\sigma}_K + \|D_c\|\widetilde{\sigma}_K])(\|K_cC_c\| + [\|K_c\|\widetilde{\sigma}_C + \widetilde{\sigma}_C\widetilde{\sigma}_K + \|C_c\|\widetilde{\sigma}_K]\,) \end{aligned} \tag{A24}$$

which also yields that $\|\widetilde{A}_{dc}\| \leq \varepsilon\widetilde{a}_{dc} + o(\varepsilon)$ *in the case that* $\varepsilon < 1$ *after grouping all the additive contributions of terms in* ε^i *for* $i \geq 2$ *into an additive bounded term, which converges to zero as* $\varepsilon \to 0$, *where*

$$\begin{aligned} \widetilde{a}_{dc} &= (1-\|K_cD_c\|)^{-1} \\ &\times[\widetilde{\sigma}_B\|K_cC_c\| + (\,\|B_c\| + \|K_cC_c\|)(\|K_c\|\widetilde{\sigma}_C + \|C_c\|\widetilde{\sigma}_K + C\,[\|K_c\|\widetilde{\sigma}_D + \|D_c\|\widetilde{\sigma}_K])] \end{aligned} \tag{A25}$$

□

Now, the following technical result follows directly from Lemma B2 and Theorem A1 (i):

Lemma B3. *Define* $\overline{\varepsilon}_2 = 1/\left(\widetilde{a}_{dc}\sup_{\omega\in \mathbf{R}_{0+}}\|(i\omega I_n - A_{cc})^{-1}\|_2\right)$ *from Equation (A25) and assume that* $\|K_cD_c\|_2 < 1$ *and that* $\overline{\varepsilon}_2 < \min(\overline{\varepsilon}, \overline{\varepsilon}_1, 1)$. *Then,* A_{dc} *is stable if* A_{cc} *is stable and* $\varepsilon \in [0, \overline{\varepsilon}_2]$. *By using Equations (A22) and (A23), a better bound of the maximum allowable* $\|A_{dc}\|_2$ *is found as* $\varepsilon \in [0, \overline{\varepsilon}_3]$ *with* $\overline{\varepsilon}_3 < \min(\overline{\varepsilon}, \overline{\varepsilon}_{20}, 1)$ *and* $\overline{\varepsilon}_{20} = \frac{\sqrt{u^2+4v\omega}-u}{2v}$.

Proof. Note from Theorem A1 (i) that the H_∞-norm of $(sI_n - A_{cc})^{-1}\widetilde{A}_{dc}$ satisfies $\|(sI_n - A_{cc})^{-1}\widetilde{A}_{dc}\|_\infty < 1$, which is guaranteed if $\varepsilon < \overline{\varepsilon}_2$, then A_{dc} is stable since A_{cc} is stable. The result follows by taking also into account, in addition, the constraints in Equations (A15) and (A16) of Lemma B.1 by using $\varepsilon^3 \leq \varepsilon^2 \leq \varepsilon$. If the second power of ε is considered and the third one is upper-bounded as $\varepsilon^3 \leq \varepsilon^2$, we examine the stability constraint $u\varepsilon + v\varepsilon^2 < \omega = 1/\sup_{\omega\in \mathbf{R}_{0+}}\|(i\omega I_n - A_{cc})^{-1}\|_2$ by building the convex parabola $\theta(\varepsilon) = v\varepsilon^2 + u\varepsilon - \omega < 0$ whose negative and positive zeros are $\varepsilon_{1,2} = \frac{-u\pm\sqrt{u^2+4v\omega}}{2v}$. Hence, the second part of the result. □

References

1. Ibeas, A.; de la Sen, M. Robustly stable adaptive control of a tandem of master-slave robotic manipulators with force reflection by using a multiestimation scheme. *IEEE Trans. Cybern. Part B Cybern.* **2006**, *36*, 1162–1164. [CrossRef]
2. Barambones, O.; Garrido, A.J.; Garrido, I. Robust speed estimation and control of an induction motor drive based on artificial neural networks. *Int. J. Adapt. Control Signal Process.* **2008**, *22*, 440–444. [CrossRef]
3. Bakule, L.; de la Sen, M. Decentralized stabilization of networked complex composite systems with nonlinear perturbations. In Proceedings of the 2009 International Conference on Control and Automation, Christchurch, New Zealand, 9–11 December 2009; Volume 1–3, pp. 2272–2277.
4. Singh, M.G. *Decentralised Control*; North-Holland Systems and Control Series; North Holland Publishing Company: New York, NY, USA, 1981; Volume 1.
5. Berman, A.; Plemmons, R.J. *Nonnegative Matrices in the Mathematical Sciences*; Academic Press: New York, NY, USA, 1979.
6. Kailath, T. *Linear Systems*; Prentice-Hall Inc.: Englewood Cliffs, NJ, USA, 1980.
7. Ortega, J.M. *Numerical Analysis*; Academic Press: New York, NY, USA, 1972.

8. Shi, H.; Li, P.; Wang, L.; Su, C.; Yu, J.; Cao, J. Delay-range-dependent robust constrained model predictive control for industrial processes with uncertainties and unknown disturbances. *Complexity* **2019**, *2019*, 2152014. [CrossRef]
9. Rosas-Jaimes, O.A.; Munoz-Hernandez, G.A.; Mino-Aguilar, G.; Castaneda-Camacho, J.; Gracios-Marin, C.A. Evaluating fractional PID control in a nonlinear MIMO model of a hydroelectric power station. *Complexity* **2019**, *2019*, 9367291. [CrossRef]
10. Xu, L.; Lou, S.; Xu, P.; Zhang, G. Feedback control and parameter invasion for a discrete competitive Lotka-Volterra system. *Discret. Dyn. Nat. Soc.* **2018**, *2018*, 7473208. [CrossRef]
11. Rana, S.M.S. Chaotic dynamics and control of discrete ratio-dependent predator-prey system. *Discret. Dyn. Nat. Soc.* **2017**, *2017*, 4537450. [CrossRef]
12. Hassan, A.; Dvorkin, Y.; Deka, D.; Chertkov, M. Chance-constrained ADMM approach for decentralized control of distributed energy resources. In Proceedings of the 2018 Power Systems Computation Conference (PSCC), Dublin, Ireland, 11–15 June 2018; pp. 1–7.
13. Shao, S.; Yang, H.; Jiang, B.; Cheng, S. Decentralized fault tolerant control for a class of interconnected nonlinear systems. *IEEE Trans. Cybern.* **2018**, *48*, 178–186. [CrossRef] [PubMed]
14. Nayyar, N.; Kalathil, D.; Jain, R. Optimal decentralized control with asymmetric one-step delayed information sharing. *IEEE Trans. Control Netw. Syst.* **2018**, *5*, 653–654. [CrossRef]
15. Wei, C.; Luo, J.; Yin, Z.; Wei, X.; Yuan, J. Robust estimation-free decentralized prescribed performance control of nonaffine nonlinear large-scale systems. *Int. J. Robust Nonlinear Control* **2018**, *28*, 174–196. [CrossRef]
16. Wang, Y.; Lopez, J.A.; Sznaier, M. Convex optimization approaches to information structured decentralized control. *IEEE Trans. Autom. Control* **2018**, *63*, 3393–3403. [CrossRef]
17. Zhang, H.G.; Yu, H.; Hao, F. Decentralized integral-based event-triggered stabilization for linear plant with actuator saturation and output feedback. *Appl. Sci.* **2017**, *7*, 11. [CrossRef]
18. Lotfy, M.E.; Senjyu, T.; Farahat, M.A.F.; Abdel-Gawad, A.F.; Lei, L.; Datta, M. Hybrid genetic algorithm fuzzy-based control schemes for small power system with high penetration wind farms. *Appl. Sci.* **2018**, *8*, 373. [CrossRef]
19. Fu, Q.; Gu, P.; Wu, J.-R. Decentralized Iterative Learning Control for Large-scale Interconnected Linear Systems with Fixed Initial Shifts. *Int. J. Control. Syst.* **2017**, *15*, 1991–1994. [CrossRef]
20. He, W.; Dong, Y. Adaptive fuzzy neural network control for a constrained robot using impedance learning. *IEEE Trans. Neural Netw. Learn. Syst.* **2018**, *29*, 1174–1186. [CrossRef] [PubMed]
21. He, W.; Dong, Y. Design and adaptive control for an upper limb robotic exoskeleton in presence of input saturation. *IEEE Trans. Neural Netw. Learn. Syst.* **2019**, *30*, 97–108. [CrossRef] [PubMed]

applied sciences

MDPI

Review

Short Overview of Early Developments of the Hardy Cross Type Methods for Computation of Flow Distribution in Pipe Networks

Dejan Brkić [1,2,3,*,†] and Pavel Praks [1,2,*]

1 European Commission, Joint Research Centre (JRC), 21027 Ispra, Italy
2 IT4Innovations, VŠB—Technical University of Ostrava, 708 00 Ostrava, Czech Republic
3 Research and Development Center "Alfatec", 18000 Niš, Serbia
* Correspondence: dejanbrkic0611@gmail.com (D.B.); Pavel.Praks@vsb.cz or Pavel.Praks@gmail.com (P.P.)
† On a short visit at IT4Innovations, VŠB—Technical University of Ostrava.

Received: 9 March 2019; Accepted: 16 April 2019; Published: 16 May 2019

Abstract: Hardy Cross originally proposed a method for analysis of flow in networks of conduits or conductors in 1936. His method was the first really useful engineering method in the field of pipe network calculation. Only electrical analogs of hydraulic networks were used before the Hardy Cross method. A problem with flow resistance versus electrical resistance makes these electrical analog methods obsolete. The method by Hardy Cross is taught extensively at faculties, and it remains an important tool for the analysis of looped pipe systems. Engineers today mostly use a modified Hardy Cross method that considers the whole looped network of pipes simultaneously (use of these methods without computers is practically impossible). A method from a Russian practice published during the 1930s, which is similar to the Hardy Cross method, is described, too. Some notes from the work of Hardy Cross are also presented. Finally, an improved version of the Hardy Cross method, which significantly reduces the number of iterations, is presented and discussed. We also tested multi-point iterative methods, which can be used as a substitution for the Newton–Raphson approach used by Hardy Cross, but in this case this approach did not reduce the number of iterations. Although many new models have been developed since the time of Hardy Cross, the main purpose of this paper is to illustrate the very beginning of modeling of gas and water pipe networks and ventilation systems. As a novelty, a new multi-point iterative solver is introduced and compared with the standard Newton–Raphson iterative method.

Keywords: Hardy Cross method; pipe networks; piping systems; hydraulic networks; gas distribution

1. Introduction

Hardy Cross solved the problem of distribution of flow in networks of pipes in his article "Analysis of Flow in Networks of Conduits or Conductors" [1] published on 13 November 1936.

Networks of pipes are nonlinear systems since the relation between flow and pressure is not linear. On the contrary, the relation between current and voltage in electrical networks with regular resistors is governed by the linear Ohm's law. Electrical circuits with diodes as well as hydraulic networks are nonlinear systems where resistance depends on current and voltage, i.e., on flow and pressure, respectively [2]. Nonlinear electrical circuits are electrical circuits containing nonlinear components. Nonlinear components can be resistive, capacitive, and inductive.

The distribution of flow in a network of pipes depends on the known inputs and consumptions at all nodes, on the given geometry of pipes, and on network topology. A stable state of flow in a network must satisfy Kirchhoff's laws, which are statements of the conservation of mass and energy. Although in theory an indefinite number of flow distributions that satisfy that the conservation of mass is possible, only one distribution from this set also satisfies the conservation of energy for all closed paths formed by pipes in the network. This state is unique for the given network and in- and outflows [3].

Since the relation between flow and pressure is not linear, Hardy Cross used a relation between an increment of flow and an increment of pressure, as this relation is linear for the given quantity of flow. If, however, the increments are fairly large, this linear relation is somewhat in error, such as for gas compressible flow. However, if the pressure drop in pipes is minor, such as in a municipality network for natural gas distribution, the Hardy Cross method can be used without significant errors [4–6]. Moreover, the Hardy Cross method can also be used for water pipe networks (district heating [7] and cooling networks [8]) and ventilation systems [9,10] (a related formulation is presented in Appendix).

The Hardy Cross method is an iterative method, i.e., a method using successive corrections [4]. Lobačev and Andrijašev in the 1930s, writing in Russian, offered similar methods [11,12]. Probably because of the language barrier and the political situation in Soviet Russia, Hardy Cross was not aware of Lobačev and Andrijašev's contributions.

Today, engineers use the most improved version of the Hardy Cross method (the ΔQ method [13]; for Δp, see [14]), which analyzes the whole looped network of pipes simultaneously [15].

As a novel approach presented for the first time here, we tested multi-point iterative methods [16,17] that can be used as a substitution for the Newton–Raphson approach used by Hardy Cross. This approach, however, did not in this case reduce the number of required iterations to reach the final balanced solution.

One example of the pipe network for distribution of gas is analyzed using the original Hardy Cross method [1] in Section 3.1, its related equivalent from Russian literature [11,12] in Section 3.2, the improved version of the Hardy Cross method [15,17,18] in Section 3.3, and finally the approach which uses multi-point iterative methods instead of the commonly used Newton–Raphson method in Section 3.4.

2. Network Piping System; Flow Distribution Calculation

2.1. Topology of the Network

The first step in solving a pipe network problem is to make a network map showing pipe diameters, lengths and connections between pipes (nodes). Sources of natural gas supply and consumption rates have to be assigned to nodes. For convenience in locating pipes, code numbers are assigned to each pipe and closed loop of pipes (represented by roman numbers for loops in Figure 1). Pipes on the network periphery are common to one loop and those in the network interior are common to two loops. Figure 1 is an example of a pipe network for distribution of natural gas for consumption in households.

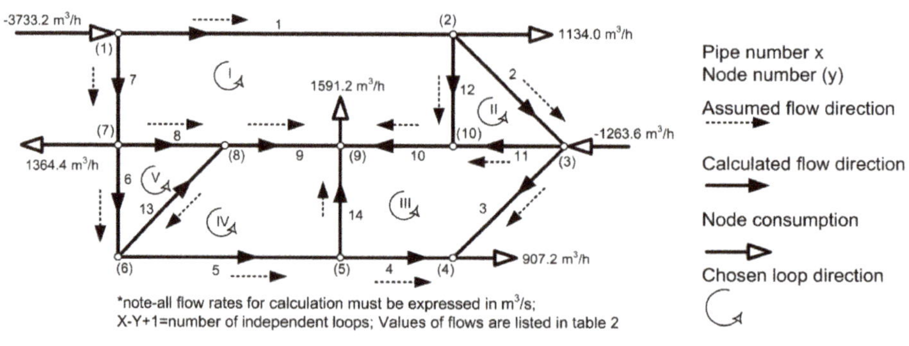

Figure 1. The network of pipes for natural gas distribution for domestic consumption.

The next step is to write the initial gas flow distribution through pipes in the network. This distribution must be done according to Kirchhoff's first law. The choice of initial flows is not critical, and the criterion should satisfy Kirchhoff's first law for every node in the network [3]. The total gas flow arriving at a node equals the total gas flow that leaves that node. The same conservation law is also valid for the whole network in total (except for gas input and output nodes that cannot be changed during calculations; see consumption nodes in Figure 1). The sum of pseudo-pressure drops along any closed path must be approximately zero for the network to be in balance according to Kirchhoff's second law. In this paper, the flow distribution, which satisfies both of Kirchhoff's laws, will be calculated using the Hardy Cross iterative method.

2.2. A Hydraulic Model

The Renouard formula; Equation (1) best fits a natural gas distribution system built with polyvinyl chloride (PVC) pipes [19,20]. Computed pressure drops are always less than the actual drop since the maximal consumption occurs only during extremely severe winter days [21,22].

$$f = \Delta\widetilde{p}^2 = p_1^2 - p_2^2 = 4810 \cdot \frac{\rho_r \cdot L \cdot Q^{1.82}}{D^{4.82}} \quad (1)$$

where f is a function of pressure, ϱ_r is relative gas density (dimensionless), $\rho_r = 0.64$, L is the pipe length (m), D is the pipe diameter (m), Q is flow (m^3/s), and p is pressure (Pa).

As shown in Appendix A, other formulas are used in the case of waterworks systems [23,24] and ventilation networks [7].

Regarding the Renouard formula (Equation (1)), one has to be careful since the pressure drop function, f, does not relate pressure drop, but actually the difference of the quadratic pressure at the input and the output of the pipe. This means that $\sqrt{\Delta\widetilde{p}^2} = \sqrt{p_1^2 - p_2^2}$ is not actually pressure drop despite using the same unit of measurement, i.e., the same unit is used as for pressure (Pa). The parameter $\sqrt{\Delta\widetilde{p}^2}$ can be noted as a pseudo-pressure drop. In fact, the gas is actually compressed, and hence that volume of the gas is decreased, and then such a compressed volume of the gas is conveying with a constant density through the gas distribution pipeline. The operating pressure for a typical distribution gas network is 4×10^5 Pa abs, i.e., 3×10^5 Pa gauge, and accordingly the volume of the gas decreases four times compared to the volume of the gas in normal (or standard) conditions. Pressure in the Renouard formula is for normal (standard) conditions.

The first derivative f' of the Renouard relation (Equation (2)), where the flow is treated as a variable, is used in the Hardy Cross method.

$$f' = \frac{\partial f(Q)}{\partial Q} = 1.82 \cdot 4810 \cdot \frac{\rho_r \cdot L \cdot Q^{0.82}}{D^{4.82}} \quad (2)$$

First assumed gas flow in each pipe is listed in the third column of Table 1. The plus or minus sign preceding flow indicates the direction of flow through the pipe for the particular loop [18,25]. A plus sign denotes counterclockwise flow in the pipe within the loop, while the minus sign clockwise. The loop direction can be chosen to be clockwise or counterclockwise (in Figure 1, all loops are counterclockwise).

3. The Hardy Cross Method; Different Versions

The Hardy Cross method is presented here: the original approach in Section 3.1, a version of the Hardy Cross method from Russian practice in Section 3.2, the modified Hardy Cross method in Section 3.3, and finally the method that uses multi-point iterative procedures instead the Newton–Raphson method and which can be implemented in all the aforementioned methods.

3.1. The Hardy Cross Method; Original Approach

The pressure drop function for each pipe is listed in Table 1 (for initial flow pattern, in the fourth column). The sign in front of the pressure drop function shown in the fourth column is the same as for flow from the observed iteration. The fifth column of Table 1 includes the first derivatives of the pressure drop function, where the flow is treated as a variable. The column of the function of pressure drops is computed algebraically, while the column of the first derivatives is estimated numerically for each loop. Flow correction ΔQ has to be computed for each loop x (Equation (3)).

$$\Delta Q_x = \left(\frac{\sum \pm f}{\left|f'\right|}\right)_x = \left(\frac{\sum \pm 4810 \cdot \frac{\rho_r \cdot L \cdot Q^{1.82}}{D^{4.82}}}{\sum \left|1.82 \cdot 4810 \cdot \frac{\rho_r \cdot L \cdot Q^{0.82}}{D^{4.82}}\right|}\right)_x \quad (3)$$

For the network from Figure 1, flow corrections for the first iteration in each loop can be calculated using Equation (4).

$$\left.\begin{array}{c} \left(\left|-f_1'\right| + \left|f_7'\right| + \left|f_8'\right| + \left|f_9'\right| + \left|-f_{10}'\right| + \left|-f_{12}'\right|\right) \cdot \Delta Q_I = -f_1 + f_7 + f_8 + f_9 - f_{10} - f_{12} \\ \left(\left|-f_2'\right| + \left|-f_{11}'\right| + \left|f_{12}'\right|\right) \cdot \Delta Q_{II} = -f_2 - f_{11} + f_{12} \\ \left(\left|-f_3'\right| + \left|f_4'\right| + \left|f_{10}'\right| + \left|f_{11}'\right| + \left|-f_{14}'\right|\right) \cdot \Delta Q_{III} = -f_3 + f_4 + f_{10} + f_{11} - f_{14} \\ \left(\left|f_5'\right| + \left|-f_9'\right| + \left|f_{13}'\right| + \left|f_{14}'\right|\right) \cdot \Delta Q_{IV} = f_5 - f_9 + f_{13} + f_{14} \\ \left(\left|f_6'\right| + \left|-f_8'\right| + \left|f_{13}'\right|\right) \cdot \Delta Q_V = f_6 - f_8 + f_{13} \end{array}\right\} \quad (4)$$

In the second iteration, the calculated correction ΔQ has to be added algebraically to the assumed gas flow (the first initial flow pattern). Further, the calculated correction ΔQ has to be subtracted algebraically from the gas flow computed in the previous iteration. This means that the algebraic operation for the first correction is the opposite of its sign, i.e., add when the sign is minus, and vice versa. A pipe common to two loops receives two corrections simultaneously. The first correction is from the particular loop under consideration, while the second one is from the adjacent loop, which the observed pipe also belongs to.

Table 1. Procedure for the solution of the flow problem for the network from Figure 1 using the modified Hardy Cross method (first two iterations)—First iteration.

Iteration 1							
Loop	**Pipe**	[a] Q	[b] $f = p_1^2 - p_2^2$	[c] $\lvert f'\rvert$	[d] ΔQ_1	[e] ΔQ_2	[f] $Q_1 = Q$
I	1	−0.3342	−144518566.8	787025109.2	−0.0994		−0.4336
	7	+0.7028	+859927106.7	2226902866.0	−0.0994		+0.6034
	8	+0.3056	+306964191.0	1828124435.8	−0.0994	−0.0532 =	+0.1530
	9	+0.2778	+800657172.4	5245486154.8	−0.0994	−0.0338 =	+0.1446
	10	−0.1364	−241342976.1	3220265516.7	−0.0994	+0.0142 ‡	−0.2217
	12	−0.0167	−6238747.4	679911398.4	−0.0994	+0.0651 ‡	−0.0511
	Σ		f_I = +1575448179.8	13987715480.9			
II	2	−0.0026	−80628.9	56440212.4	−0.0651		−0.0677
	11	−0.1198	−14582531.0	221537615.9	−0.0651	+0.0142 ‡	−0.1707
	12	+0.0167	+6238747.4	679911398.4	−0.0651	+0.0994 ∓	+0.0511
	Σ		f_{II} = −8424412.4	957889226.7			
III	3	−0.2338	−406110098.1	3161336093.1	−0.0142		−0.2480
	4	+0.0182	+1530938.1	153093808.5	−0.0142		+0.0040
	10	+0.1364	+241342976.1	3220265516.7	−0.0142	+0.0994 ∓	+0.2217
	11	+0.1198	+14582531.0	221537615.9	−0.0142	+0.0651 ∓	+0.1707
	14	−0.0278	−21840183.8	1429824980.5	−0.0142	−0.0338 ±	−0.0757
	Σ		f_{III} = −170493836.7	8186058014.8			
IV	5	+0.0460	+7523646.2	297674697.0	+0.0338		+0.0798
	9	−0.2778	−800657172.4	5245486154.8	+0.0338	+0.0994 ‡	−0.1446
	13	+0.0278	+21840183.8	1429824980.5	+0.0338	−0.0532 =	+0.0084
	14	+0.0278	+21840183.8	1429824980.5	+0.0338	+0.0142∓	+0.0757

Table 1. *Cont.*

Iteration 1							
Loop	Pipe	[a] Q	[b] $f=p_1^2-p_2^2$	[c] $\lvert f' \rvert$	[d] ΔQ_1	[e] ΔQ_2	[f] $Q_1 = Q$
	Σ		$f_{IV} = -749453158.7$	8402810812.8			
V	6	+0.0182	+3479197.2	347919720.0	+0.0532		+0.0714
	8	−0.3056	−306964191.0	1828124435.8	+0.0532	+0.0994 ‡	−0.1530
	13	−0.0278	−21840183.8	1429824980.5	+0.0532	−0.0338 ±	−0.0084
	Σ		$f_V = -325325177.5$	3605869136.3			
Iteration 2							
Loop	Pipe	$Q_1 = Q$	$f=p_1^2-p_2^2$	$\lvert f' \rvert$	ΔQ_1	ΔQ_2	$Q_2 = Q$
I	1	−0.4336	−232172997.6	974431560.7	−0.0058		−0.4394
	7	+0.6034	+651439280.6	1965036192.1	−0.0058		+0.5976
	8	+0.1530	+87112249.4	1036457217.8	−0.0058	−0.0178=	+0.1294
	9	+0.1446	+243990034.4	3070921097.1	−0.0058	−0.0098=	+0.1290
	10	−0.2217	−584137977.5	4795666298.0	−0.0058	+0.0018 ‡	−0.2257
	12	−0.0511	−47725420.6	1700518680.1	−0.0058	$-2.1\cdot10^{-5}$ ±	−0.0569
	Σ		$f_I = +118505168.7$	13543031045.9			
II	2	−0.0677	−30372941.9	816962908.0	$+2.1\cdot10^{-5}$		−0.0676
	11	−0.1707	−27780459.9	296182372.8	$+2.1\cdot10^{-5}$	+0.0018 ‡	−0.1689
	12	+0.0511	+47725420.6	1700518680.1	$+2.1\cdot10^{-5}$	+0.0058 ∓	+0.0569
	Σ		$f_{II} = -10427981.2$	2813663960.8			
III	3	−0.2480	−451970989.4	3317464222.8	−0.0018		−0.2497
	4	+0.0040	+99061.2	44589235.4	−0.0018		+0.0023
	10	+0.2217	+584137977.5	4795666298.0	−0.0018	+0.0058 ∓	+0.2257
	11	+0.1707	+27780459.9	296182372.8	−0.0018	$-2.1\cdot10^{-5}$ =	+0.1689
	14	−0.0757	−135261698.0	3251481942.9	−0.0018	−0.0098 ±	−0.0873
	Σ		$f_{III} = +24784811.3$	11705384072.0			
IV	5	+0.0798	+20483898.1	467437803.0	+0.0098		+0.0896
	9	−0.1446	−243990034.4	3070921097.1	+0.0098	+0.0058 ‡	−0.1290
	13	+0.0084	+2454799.0	534076127.2	+0.0098	−0.0178 =	+0.0004
	14	+0.0757	+135261698.0	3251481942.9	+0.0098	+0.0018 ∓	+0.0873
	Σ		$f_{IV} = -85789639.2$	7323916970.2			
V	6	+0.0714	+41857166.9	1067095933.1	+0.0178		+0.0892
	8	−0.1530	−87112249.4	1036457217.8	+0.0178	+0.0058 ‡	−0.1294
	13	−0.0084	−2454799.0	534076127.2	+0.0178	−0.0098 ±	−0.0004
	Σ		$f_V = -47709881.5$	2637629278.1			

[a] Pipe lengths, diameters and initial flow distribution are shown in Table 2 and Figure 1; [b] f calculated using the Renouard Equation (1); [c] f' calculated using the first derivative of the Renouard, Equation (2), where flow is variable; [d] calculated using the matrix Equation (10) and entering ΔQ with the opposite sign (using the original Hardy Cross method for Iteration 1: $\Delta Q_I = +0.1126$; $\Delta Q_{II} = -0.0088$; $\Delta Q_{III} = -0.0208$; $\Delta Q_{IV} = -0.0892$; $\Delta Q_V = -0.0902$; using the Lobačev method for Iteration 1: $\Delta Q_I = -0.1041$; $\Delta Q_{II} = -0.0644$; $\Delta Q_{III} = -0.0780$; $\Delta Q_{IV} = +0.1069$; $\Delta Q_V = -0.1824$); [e] ΔQ_2 is ΔQ_1 from the adjacent loop; [f] the final calculated flow in the first iteration is used for the calculation in the second iteration, etc.; [g] if Q and Q_1 have a different sign, this means that the flow direction is opposite to that in the previous iteration, etc. (this occurs with the flow in pipe 13 between Iteration 3 and 4).

The upper sign after the second correction in Table 1 is plus if the flow direction in the mutual pipe coincides with the assumed orientation of the adjacent loop, and minus if it does not (Figure 2). The lower sign is the sign in front of correction ΔQ calculated for the adjacent loop (Figure 2).

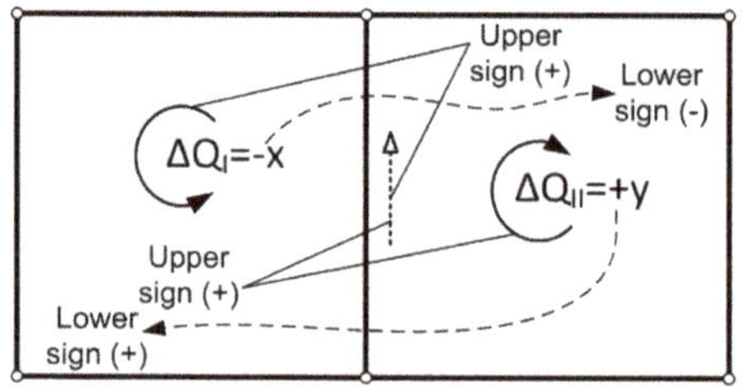

Figure 2. Rules for the upper and lower sign (correction from the adjacent loop; second correction).

Details of the signs of corrections were reported by Brkić [18] and Corfield et al. [25].

The algebraic operation for the second correction should be the opposite of its lower sign when its upper sign is the same as the sign in front of flow Q, and, as indicated by its lower sign, when its upper sign is opposite to the sign in front of flow Q.

The calculation procedure is repeated until the net algebraic sum of pressure functions around each loop is as close to zero as the desired degree of precision demands. This also means that the calculated corrections of flow and the change in calculated flow between two successive iterations is approximately zero. The pipe network is then in approximate balance and the calculation after the Hardy Cross can be terminated.

In the original Hardy Cross method, the corrections for the first iteration are:

$$\Delta Q_I = \frac{1575448179.8}{13987715480.9} = +0.1126$$

$$\Delta Q_{II} = \frac{-8424412.4}{957889226.7} = -0.0088$$

$$\Delta Q_{III} = \frac{-170493836.7}{8186058014.8} = -0.0208$$

$$\Delta Q_{IV} = \frac{-749453158.7}{8402810812.8} = -0.0892$$

and

$$\Delta Q_V = \frac{-325325177.5}{3605869136.3} = -0.0902$$

3.2. A Version of the Hardy Cross Method from Russian Practice

As mentioned in the Introduction, two Russian authors, Lobačev [11] and Andrijašev [12], proposed a similar method to Hardy Cross [1]. These two methods are also from the 1930s. It is not clear if Hardy Cross had been aware of the contribution of these two authors from Soviet Russia and vice versa, but most probably the answer to this question is no, for both sides. The main difference between the Hardy Cross and Andrijašev methods is that in the method of Andrijašev contours can be defined to include few loops. This strategy only complicates the situation, while the number of required iterations remains unchanged.

Further on, the Andrijašev method can be seen from the example in the paper of Brkić [3]. Here, the method of Lobačev is shown in more detail. In the Hardy Cross method, the influence of adjacent loops is neglected. The Lobačev method takes into consideration this influence (Equation (5)):

$$\left.\begin{array}{c}
+\left(|-f_1'|+|f_7'|+|f_8'|+|f_9'|+|-f_{10}'|+|-f_{12}'|\right)\cdot\Delta Q_I+|f_{12}'|\cdot\Delta Q_{II}+|f_{10}'|\cdot\Delta Q_{III}+|f_9'|\cdot\Delta Q_{IV}+|f_8'|\cdot\Delta Q_V=-f_1+f_7+f_8+f_9-f_{10}-f_{12}\\
|f_{12}'|\cdot\Delta Q_I-\left(|-f_2'|+|-f_{11}'|+|f_{12}'|\right)\cdot\Delta Q_{II}-|f_{11}'|\cdot\Delta Q_{III}=-f_2-f_{11}+f_{12}\\
+|f_{10}'|\cdot\Delta Q_I-|f_{11}'|\cdot\Delta Q_{II}-\left(|-f_3'|+|f_4'|+|f_{10}'|+|f_{11}'|+|-f_{14}'|\right)\cdot\Delta Q_{III}-|f_{14}'|\cdot\Delta Q_{IV}=-f_3+f_4+f_{10}+f_{11}-f_{14}\\
+|f_9'|\cdot\Delta Q_I-|f_{14}'|\cdot\Delta Q_{III}-\left(|f_5'|+|-f_9'|+|f_{13}'|+|f_{14}'|\right)\cdot\Delta Q_{IV}-|f_{13}'|\cdot\Delta Q_V=f_5-f_9+f_{13}+f_{14}\\
+|f_8'|\cdot\Delta Q_I-|f_{13}'|\cdot\Delta Q_{IV}-\left(|f_6'|+|-f_8'|+|f_{13}'|\right)\cdot\Delta Q_V=f_6-f_8+f_{13}
\end{array}\right\} \quad (5)$$

In Equation (5), signs in front of terms from the left side of the equal signs have to be determined (this is much more complex than in the Hardy Cross method). Thus, in the Lobačev method, if $(\sum f)_x > 0$, then the sign in front of $\left(\sum\left|f'\right|\right)_x$ has to be positive, and vice versa (for the first iteration, this can be seen in Table 1; $f_I = +1575448179.8 > 0$, $f_{II} = -8424412.4 < 0$, $f_{III} = -170493836.7 < 0$, $f_{IV} = -749453158.7 < 0$, $f_V = -325325177.5 < 0$). The sign for the other terms (these terms are sufficient in the Hardy Cross method) are determined using further rules and the scheme in Figure 3.

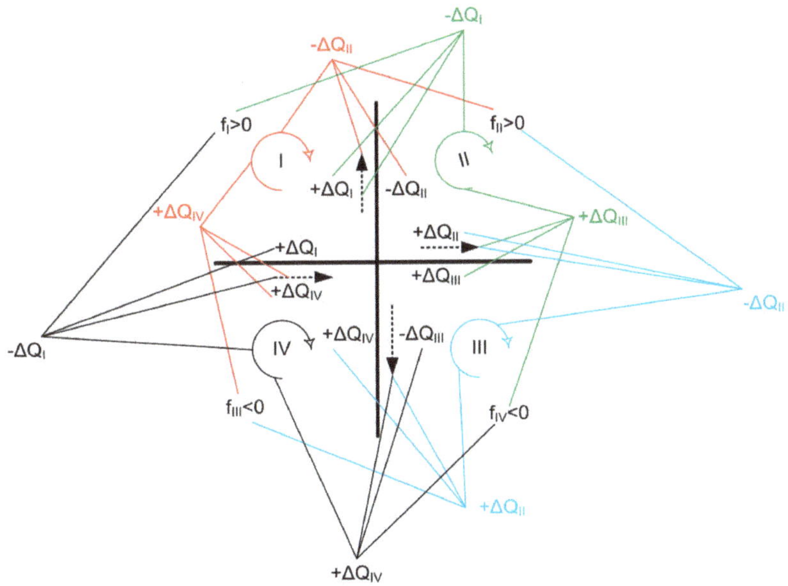

Figure 3. Rules for terms from Lobačev equations, which do not exist in the Hardy Cross method.

As shown in Figure 3 if $(\sum f)_x > 0$ and if the assumed flow coincides with the loop direction, then the sign of flow in the adjacent pipe is negative and, if the flow does not coincide with the loop direction, then the sign of flow in the adjacent pipe is positive. Conversely, if $(\sum f)_x < 0$ and if the assumed flow coincides with the loop direction, then the sign of flow in the adjacent pipe is positive and, if the flow does not coincide with the loop direction, then the sign of flow in the adjacent pipe is negative. This procedure determines the signs in the front of the flow corrections (ΔQ), which are shown in Figure 3 with black letters (and also in Figure 4 for our pipe network example).

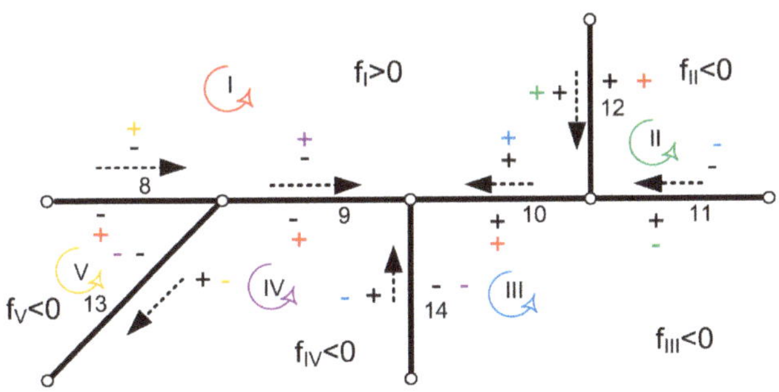

Figure 4. Rules for terms from Lobačev equations, which do not exist in the Hardy Cross method applied for the network from Figure 1.

If $(\sum f)_x$ from the adjacent loop is positive, while the loop direction and assumed flow do not coincide, the flow correction from the adjacent loop changes its sign, and, conversely, if $(\sum f)_x$ from the adjacent loop is positive, while the loop direction and assumed flow coincide, the flow correction from the adjacent loop does not change its sign. If $(\sum f)_x$ from the adjacent loop is negative, while the loop direction and assumed flow do not coincide, the flow correction from the adjacent loop does not change its sign, and, conversely, if $(\sum f)_x$ from the adjacent loop is negative, while the loop direction and assumed flow do not coincide, the flow correction from the adjacent loop changes its sign. These four parameters are connected in Figure 3 with the same colored lines. Flow corrections (ΔQ) shown in Figure 4 with different colors are used with the related signs in Equation (5). They are chosen in a similar way as explained in the example in Figure 3.

Thus, instead of the simple equations of the original Hardy Cross method, the system of equations has to be solved in the Lobačev method (Equation (6)).

$$\left.\begin{array}{c} +13987715480.9{\cdot}\Delta Q_I \underline{+ 679911398.4{\cdot}\Delta Q_{II} + 3220265516.7{\cdot}\Delta Q_{III} + 5245486154.8{\cdot}\Delta Q_{IV} + 1828124435.8{\cdot}\Delta Q_V} = +1575448179.8 \\ \underline{+679911398.4{\cdot}\Delta Q_I} - 957889226.7{\cdot}\Delta Q_{II} \underline{- 221537615.9{\cdot}\Delta Q_{III}} = -8424412.4 \\ \underline{+3220265516.7{\cdot}\Delta Q_I - 221537615.9{\cdot}\Delta Q_{II}} - 8186058014.8{\cdot}\Delta Q_{III} \underline{- 1429824980.5{\cdot}\Delta Q_{IV}} = -170493836.7 \\ \underline{+5245486154.8{\cdot}\Delta Q_I - 1429824980.5{\cdot}\Delta Q_{III}} - 8402810812.8{\cdot}\Delta Q_{IV} \underline{- 1429824980.5{\cdot}\Delta Q_V} = -749453158.7 \\ \underline{+1828124435.8{\cdot}\Delta Q_I - 1429824980.5{\cdot}\Delta Q_{IV}} - 3605869136.3{\cdot}\Delta Q_V = -325325177.5 \end{array}\right\} \quad (6)$$

Underlined terms in Equation (6) do not exist in the Hardy Cross method.

In the Lobačev method, corrections for the first iterations are $\Delta Q_x = \frac{\Delta(\Delta Q_x)}{\Delta}$, where Δ for the first iteration is (Equation (7)):

$$\Delta = \begin{vmatrix} 13987715480.9 & 679911398.4 & 3220265516.7 & 5245486154.8 & 1828124435.8 \\ 679911398.4 & -957889226.7 & -221537615.9 & 0 & 0 \\ 3220265516.7 & -221537615.9 & -8186058014.8 & -1429824980.5 & 0 \\ 5245486154.8 & 0 & -1429824980.5 & -8402810812.8 & -1429824980.5 \\ 1828124435.8 & 0 & 0 & -1429824980.5 & -3605869136.3 \end{vmatrix} = +3.97{\cdot}10^{+48} \quad (7)$$

while ΔQ_x for the first iteration is (Equation (8)).

$$\Delta(\Delta Q_I) = \begin{vmatrix} 1575448179.8 & 679911398.4 & 3220265516.7 & 5245486154.8 & 1828124435.8 \\ -8424412.4 & -957889226.7 & -221537615.9 & 0 & 0 \\ -170493836.7 & -221537615.9 & -8186058014.8 & -1429824980.5 & 0 \\ -749453158.7 & 0 & -1429824980.5 & -8402810812.8 & -1429824980.5 \\ -325325177.5 & 0 & 0 & -1429824980.5 & -3605869136.3 \end{vmatrix} = -4.14{\cdot}10^{+47} \quad (8)$$

The correction for the first loop in the first iteration is (Equation (9)).

$$\Delta Q_I = \frac{\Delta(\Delta Q_I)}{\Delta} = \frac{-4.14{\cdot}10^{+47}}{+3.97{\cdot}10^{+48}} = -0.1041 \quad (9)$$

Other corrections in the first iteration are $\Delta Q_{II} = -0.0644$, $\Delta Q_{III} = -0.0780$, $\Delta Q_{IV} = +0.1069$ and $\Delta Q_V = -0.1824$.

The Lobačev method is more complex compared to the original Hardy Cross method. However, the number of required iterations is not reduced using the Lobačev procedure compared with the original Hardy Cross procedure.

3.3. The Modified Hardy Cross Method

The Hardy Cross method can be noted in matrix form. The gas distribution network in Figure 1 has five independent loops (Equation (10)).

$$\begin{bmatrix} \Sigma|f'_I| & 0 & 0 & 0 & 0 \\ 0 & \Sigma|f'_{II}| & 0 & 0 & 0 \\ 0 & 0 & \Sigma|f'_{III}| & 0 & 0 \\ 0 & 0 & 0 & \Sigma|f'_{IV}| & 0 \\ 0 & 0 & 0 & 0 & \Sigma|f'_V| \end{bmatrix} x \begin{bmatrix} \Delta Q_I \\ \Delta Q_{II} \\ \Delta Q_{III} \\ \Delta Q_{IV} \\ \Delta Q_V \end{bmatrix} = \begin{bmatrix} \Sigma \pm f_I \\ \Sigma \pm f_{II} \\ \Sigma \pm f_{III} \\ \Sigma \pm f_{IV} \\ \Sigma \pm f_V \end{bmatrix} \quad (10)$$

Equation (4) provides for each particular loop in the network the same corrections as Equation (10) using matrix calculation. Epp and Fowler [15] improved the original Hardy Cross method [1] by replacing some of the zeroes in the non-diagonal terms of Equation (10). For example, if pipe 8 is mutual for loop I and V, the first derivative of the pressure drop function for the observed pipe, where flow treated as a variable, will be put with a negative sign in the first column and the fifth row, and also in the fifth column and the first row (Equation (11)).

$$\begin{bmatrix} \Sigma|f'_I| & -f'_{12} & -f'_{10} & -f'_9 & -f'_8 \\ -f'_{12} & \Sigma|f'_{II}| & -f'_{11} & 0 & 0 \\ -f'_{10} & -f'_{11} & \Sigma|f'_{III}| & -f'_{14} & 0 \\ -f'_9 & 0 & -f'_{14} & & -f'_{13} \\ -f'_8 & 0 & 0 & -f'_{13} & \Sigma|f'_V| \end{bmatrix} x \begin{bmatrix} \Delta Q_I \\ \Delta Q_{II} \\ \Delta Q_{III} \\ \Delta Q_{IV} \\ \Delta Q_V \end{bmatrix} = \begin{bmatrix} \Sigma \pm f_I \\ \Sigma \pm f_{II} \\ \Sigma \pm f_{III} \\ \Sigma \pm f_{IV} \\ \Sigma \pm f_V \end{bmatrix} \quad (11)$$

In the modified Hardy Cross method, corrections for the first iterations are shown in Equation (12), and the solutions are listed in Table 1.

$$\begin{bmatrix} +13987715480.9 & -679911398.4 & -3220265516.7 & -5245486154.8 & -1828124435.8 \\ -679911398.4 & +957889226.7 & -221537615.9 & 0 & 0 \\ -3220265516.7 & -221537615.9 & +8186058014.8 & -1429824980.5 & 0 \\ -5245486154.8 & 0 & -1429824980.5 & +8402810812.8 & -1429824980.5 \\ -1828124435.8 & 0 & 0 & -1429824980.5 & +3605869136.3 \end{bmatrix} x \begin{bmatrix} \Delta Q_I \\ \Delta Q_{II} \\ \Delta Q_{III} \\ \Delta Q_{IV} \\ \Delta Q_V \end{bmatrix} = \begin{bmatrix} +1575448179.8 \\ -8424412.4 \\ -170493836.7 \\ -749453158.7 \\ -325325177.5 \end{bmatrix} \quad (12)$$

This procedure significantly reduces the number of iterations required for the solution of the problem (Figure 5).

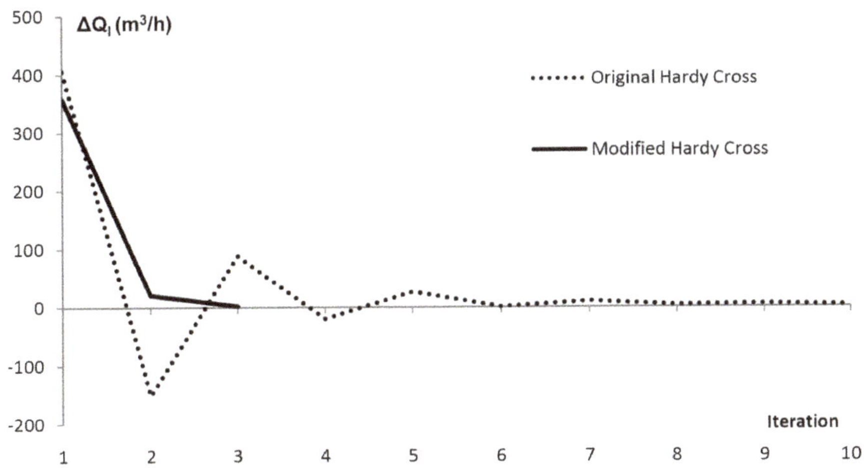

Figure 5. The number of required iterations for the solution using the original vs. the improved Hardy Cross method.

The first two iterations for the network in Figure 1 are shown in Table 1. Pipe diameters and lengths, as well as the first, assumed, and the final calculated flow distributions for the network in balance are shown in Table 2.

The gas velocity in the network is small (can be up to 10–15 m/s). The network can be the subject of diameter optimization (as in [4]), which can also be done by using the Hardy Cross method (diameter correction ΔD should be calculated for known and locked flow, where the first derivative of the Renouard function has to be calculated for diameter as a variable). The network should stay unchanged, even if planned gas consumption on nodes 5, 6, 8 and 10 increases, as pipes 4 and 13 will be useful thanks to increased gas flow.

Similar examples, but for water flow, can be seen in [26]. Optimization of pipe diameters in a water distributive pipe network using the same approach can be seen in [6].

Table 2. Pipe diameters and lengths, flows, and velocities of gas within pipes.

[a] Pipe Number	Diameter (m)	Length (m)	[b] Assumed Flows (m^3/h)	[c] Calculated Flows (m^3/h)	Gas Velocity (m/s)
1	0.305	1127.8	1203.2	1583.6	1.5
2	0.203	609.6	9.2	245.2	0.5
3	0.203	853.4	841.6	899.7	1.9
4	0.203	335.3	65.6	7.5	0.01
5	0.203	304.8	165.6	320.2	0.7
6	0.203	762.0	65.6	322.7	0.7
7	0.203	243.8	2530.0	2149.6	4.6
8	0.203	396.2	1100.0	462.4	1.0
9	0.152	304.8	1000.0	465.0	1.8
10	0.152	335.3	491.2	813.5	3.1
11	0.254	304.8	431.2	609.1	0.8
12	0.152	396.2	60.0	204.8	0.8
13	0.152	548.6	100.0	[d] −2.6	−0.009
14	0.152	548.6	100.0	312.7	1.2

[a] Network from Figure 1 (flows are for normal pressure conditions; real pressure in the network is 4×10^5 Pa abs, i.e., 3×10^5 Pa); [b] chosen to satisfy Kirchhoff's first law for all nodes (dash arrows in Figure 1); [c] calculated to satisfy Kirchhoff's first law for all nodes and Kirchhoff's second law for all closed path formed by pipes (full errors in Figure 1); [d] the minus sign means that the direction of flow is opposite to the initial pattern for assumed flows.

3.4. The Multi-Point Iterative Hardy Cross Method

The here described multipoint method can substitute the Newton–Raphson iterative procedure used in all the above described methods. Recently, we successfully used this multipoint method for acceleration of the iterative solution of the Colebrook equation for flow friction modeling [16,17]. On the contrary, for the gas network example in Figure 1, the multipoint method requires the same number of iterations as the original Newton–Raphson procedure.

For the test, we used the three-point method from Džunić et al. [27]. Flow corrections ΔQ_I from Equations (10) and (11) from the first loop I should be calculated using the three-point procedure (Equation (13)):

$$\begin{cases} \Delta Q_I = -\frac{(\Sigma \pm f_I)_i}{(\Sigma|f_I'|)_i}, \text{ where } i = 1; \\ \Delta Q_I = -\frac{(\Sigma \pm f_I)_i}{(\Sigma \pm f_I)_i - 2\cdot(\Sigma \pm f_I)_{i+1}} \cdot \frac{(\Sigma \pm f_I)_{i+1}}{(\Sigma|f_I'|)_i}, \text{ where } i = 2; \\ \Delta Q_I = -\frac{(\Sigma \pm f_x)_{i+2}}{(\Sigma|f_I'|)_i \cdot \left[1 - 2\cdot\frac{(\Sigma \pm f_I)_{i+1}}{(\Sigma \pm f_I)_i} - \left(\frac{(\Sigma \pm f_I)_{i+1}}{(\Sigma \pm f_I)_i}\right)^2\right] \cdot \left[1 - \frac{(\Sigma \pm f_I)_{i+2}}{(\Sigma \pm f_I)_{i+1}}\right] \cdot \left[1 - 2\cdot\frac{(\Sigma \pm f_I)_{i+2}}{(\Sigma \pm f_I)_i}\right]}, \text{ where } i > 2; \end{cases} \tag{13}$$

Formulas of flow corrections ΔQ_I depend on the counter i. The algorithm starts from $i = 1$, in which the multipoint method is the same as the original Newton–Raphson procedure (Equation (13a)):

$$\Delta Q_I = -\frac{(\Sigma \pm f_I)_i}{(\Sigma|f_I'|)_i} \tag{13a}$$

In the second iteration, $i = 2$, flow corrections ΔQ_I have a little bit more complicated form (Equation (13b)):

$$\Delta Q_I = -\frac{(\sum \pm f_I)_i}{(\sum \pm f_I)_i - 2\cdot(\sum \pm f_I)_{i+1}} \cdot \frac{(\sum \pm f_I)_{i+1}}{\left(\sum\left|f_I'\right|\right)_i} \quad (13b)$$

The symbol $(\sum \pm f_I)_i$ represents stored values from the first iteration, whereas $(\sum \pm f_I)_{i+1}$ represents values from the second iteration.

For the third iteration, $i = 3$, flow corrections ΔQ_I have the most complicated form (Equation (13c)):

$$\Delta Q_I = -\frac{(\sum \pm f_x)_{i+2}}{\left(\sum\left|f_I'\right|\right)_i \cdot \left[1 - 2\cdot\frac{(\sum \pm f_I)_{i+1}}{(\sum \pm f_I)_i} - \left(\frac{(\sum \pm f_I)_{i+1}}{(\sum \pm f_I)_i}\right)^2\right] \cdot \left[1 - \frac{(\sum \pm f_I)_{i+2}}{(\sum \pm f_I)_{i+1}}\right] \cdot \left[1 - 2\cdot\frac{(\sum \pm f_I)_{i+2}}{(\sum \pm f_I)_i}\right]} \quad (13c)$$

This iterative process can continue, as the formula from the third iteration is also used for iterations $i = 4, 5, 6, 7$, etc. This procedure should be done for all loops in the network separately (in our case for I, II, III, IV and V). However, to simplify calculations, derivative-free methods can be used [28,29].

4. Conclusions

Hardy Cross simplified mathematical modeling of complex problems in structural and hydraulic engineering long before the computer age. Moment distributions in indeterminate concrete structures described with differential equations were too complex for the time before computers. Hardy Cross later applied these finding from structural analysis to balancing of flow in pipe networks. He revolutionized how the profession addressed complicated problems. Today, in engineering practice, the modified Hardy Cross method proposed by Epp and Fowler [15] is used rather than the original version of the Hardy Cross method [1]. Methods proposed by Hamam and Brameller [30], as well as by Wood and Charles [31] and Wood and Rayes [32], are used in common practice [33], too. Moreover, the node-oriented method proposed by Shamir and Howard [34] is also based on the Hardy Cross method.

Professional engineers use a different kind of looped pipeline in professional software [35], but. even today, engineers invoke the name of Hardy Cross with awe. When petroleum and natural gas or civil engineers have to figure out what is happening in looped piping systems [36], they inevitably turn to what is generally known as the Hardy Cross method. The original Hardy Cross method is still extensively used for teaching and learning purpose [6]. Here, we introduced into the Hardy Cross method the multi-point iterative approach instead of the Newton–Raphson iterative approach, but it does not affect the number of required iterations to reach the final solution in our case.

The view of Hardy Cross was that engineers lived in the real world with real problems and that it was their job to come up with answers to questions in design tasks, even if initial approximations were involved. After Hardy Cross, the essential idea which he wished to present involves no mathematical relations except the simplest arithmetic.

For example, ruptures of pipes with leakage can be detected using the Hardy Cross method because every single-point disturbance affects the general distribution of flow and pressure [37,38].

This paper has the purpose of illustrating the very beginning of modeling of gas or water pipe networks. As noted by Todini and Rossman [39], many new models have been developed since the time of Hardy Cross.

Some details about the life and work of Hardy Cross are given in Appendix B.

Author Contributions: The paper is a product of the joint efforts of the authors, who worked together on models of natural gas distribution networks. P.P. has a scientific background in applied mathematics and programming while D.B. has a background in control and applied computing in mechanical and petroleum engineering. D.B. performed calculations with advice from P.P. who has extensive experience with the implementation of probabilistic gas network modeling tools.

Funding: Resources to cover the Article Processing Charge were provided by the European Commission. This article is registered as Pubsy JRC116624 in the internal system for publications of the European Commission, Joint Research Center.

Acknowledgments: We acknowledge support from the European Commission, Joint Research Centre (JRC), Directorate C: Energy, Transport and Climate, Unit C3: Energy Security, Distribution and Markets, and we especially thank Marcelo Masera and Ricardo Bolado-Lavin for their scientific supervision and approvals. We also thank John Cawley from IT4Innovations, who as a native speaker kindly checked the correctness of English expressions throughout the paper. This work was partially supported by the Ministry of Education, Science and Technological Development of the Republic of Serbia through the project iii44006 and by the Ministry of Education, Youth and Sports of the Czech Republic through the National Programme of Sustainability (NPS II) project "IT4Innovations excellence in science-LQ1602". Dejan Brkić acknowledges support from the mentioned Czech's project LQ1602 for funding of his stay in Ostrava at IT4Innovations.

Conflicts of Interest: The authors declare no conflict of interest. Neither the European Commission, the VŠB—Technical University of Ostrava, or the Alfatec nor any person acting on their behalf is responsible for any use which might be made of this publication.

Nomenclature

The following symbols are used in this paper:

ρ_r	relative gas density (-); here $\rho_r = 0.64$
ρ	density of air (kg/m^3); here ρ = 1.2 kg/m^3
L	length of pipe (m)
D	diameter of pipe (m)
Q	flow (m^3/s)
ΔQ	flow correction (m^3/s)
p	pressure (Pa)
Δp	pressure correction (Pa)
f	function of pressure
f'	first derivative of function of pressure
λ	Darcy (Moody) flow friction factor (dimensionless)
Re	Reynolds number (dimensionless)
$\frac{\varepsilon}{D}$	relative roughness of inner pipe surface (dimensionless)
C_d	flow discharge coefficient (dimensionless)
A	area of ventilation opening (m^2)
π	Ludolph number; π ≈3.1415
i	counter

Appendix A Hydraulic Models for Water Pipe Networks and for Ventilation Systems

To relate pressure p [40] with flow Q, instead of Equation (1), which is used for gas distribution networks in municipalities, for water distribution the Darcy–Weisbach correlation and Colebrook equation are recommended (Equation (A1)) [23,41–47], and for ventilation systems, the Atkinson equation (Equation (A2)) [9]:

$$\left.\begin{aligned} \frac{1}{\sqrt{\lambda}} &= -2\cdot log_{10}\left(\frac{2.51}{Re}\cdot\frac{1}{\sqrt{\lambda}} + \frac{\varepsilon}{3.71\cdot D}\right) \\ \Delta p &= \frac{8\cdot\rho\cdot\lambda\cdot L\cdot Q^2}{\pi^2\cdot D^5} \end{aligned}\right\} \quad \text{(A1)}$$

$$\Delta p = \frac{\rho}{2\cdot C_d^2\cdot A^2}\cdot Q^2 \quad \text{(A2)}$$

Appendix B The Life and Work of Hardy Cross

Hardy Cross (1885–1959) was one of America's most brilliant engineers [48–55]. He received a BSc degree in arts in 1902 and BSc degree in science in 1903, both from Hampden-Sydney College, where he taught English and Mathematics. Hardy Cross was also awarded a BSc degree in 1908 from Massachusetts Institute of Technology and an MCE degree from Harvard University in 1911, both in civil engineering. He taught civil engineering at Brown University from 1911 to 1918. He left teaching twice to become involved in the practice of structural and hydraulic engineering, from 1908 to 1909, and from 1918 to 1921. The most creative years of Hardy Cross were spent at the University of Illinois in Champaign-Urbana where he was a professor of structural engineering from 1921 to 1937. His famous article "Analysis of flow in networks of conduits or conductors" was published in 1936 in Urbana Champaign University Illinois Bulletin; Engineering Experiment Station number 286 [1]. His name is also famous in the field of structural engineering [53–55]. He developed the moment distribution method for statically indeterminate structures in 1932 [56]. This method has been superseded by more powerful procedures, however the moment distribution method made possible the efficient and safe design of many reinforced concrete buildings for the duration of an entire generation. Furthermore, the solution of the here discussed pipe network problems was a byproduct of his explorations in structural analysis. Later, Hardy Cross was Chair of the Department of Civil Engineering at Yale, from 1937 to the early 1950s.

Related to the moment distribution method for statically indeterminate structures developed by Hardy Cross in 1932 [56], it need to be noted that in 1922 and 1923, Konstantin A. Čališev, emigrant from Soviet Russia, writing in Serbian, offered a similar method of solving the slope deflection equations by successive approximations [57–60]. The method of Hardy Cross is an improved version of the Čališev's method, but with important circumstance that Hardy Cross most probably was not aware of Čališev's contributions. As noted in [49]: "It was Hardy Cross's genius that he recognized he could bypass adjusting rotations to get to the moment balance at each and every node.", which was the part that Čališev did not developed.

References

1. Cross, H. Analysis of flow in networks of conduits or conductors. *Eng. Exp. Stn.* **1936**, *34*, 3–29. Available online: http://hdl.handle.net/2142/4433 (accessed on 28 March 2019).
2. Katzenelson, J. An algorithm for solving nonlinear resistor networks. *Bell Syst. Tech. J.* **1965**, *44*, 1605–1620. [CrossRef]
3. Gay, B.; Middleton, P. The solution of pipe network problems. *Chem. Eng. Sci.* **1971**, *26*, 109–123. [CrossRef]
4. Brkić, D. Iterative methods for looped network pipeline calculation. *Water. Resour. Manag.* **2011**, *25*, 2951–2987. [CrossRef]
5. Brkić, D. A gas distribution network hydraulic problem from practice. *Petrol. Sci. Technol.* **2011**, *29*, 366–377. [CrossRef]
6. Brkić, D. Spreadsheet-based pipe networks analysis for teaching and learning purpose. *Spreadsheets Educ. (EJSIE)* **2016**, *9*, 4646. Available online: https://sie.scholasticahq.com/article/4646.pdf (accessed on 28 March 2019).
7. Brkić, D.; Tanasković, T.I. Systematic approach to natural gas usage for domestic heating in urban areas. *Energy* **2008**, *33*, 1738–1753. [CrossRef]
8. Augusto, G.L.; Culaba, A.B.; Tanhueco, R.M. Pipe sizing of district cooling distribution network using implicit Colebrook-White Equation. *J. Adv. Comput. Intell.* **2016**, *20*, 76–83. [CrossRef]
9. Aynsley, R.M. A resistance approach to analysis of natural ventilation airflow networks. *J. Wind Eng. Ind. Aerodyn.* **1997**, *67*, 711–719. [CrossRef]
10. Kassai, M.; Poleczky, L.; Al-Hyari, L.; Kajtar, L.; Nyers, J. Investigation of the energy recovery potentials in ventilation systems in different climates. *Facta Univ. Ser. Mech. Eng.* **2018**, *16*, 203–217. [CrossRef]
11. Лобачев, В.Г. Новый метод увязки колец при расчете водопроводных сетей. *Санитарная техника* **1934**, *2*, 8–12. (In Russian)
12. Андрияшев, М.М. Техника расчета водопроводной сети. Москва, Огиз: Советское законодательство. 1932. (In Russian)
13. Niazkar, M.; Afzali, S.H. Analysis of water distribution networks using MATLAB and Excel spreadsheet: Q-based methods. *Comput. Appl. Eng. Educ.* **2017**, *25*, 277–289. [CrossRef]
14. Niazkar, M.; Afzali, S.H. Analysis of water distribution networks using MATLAB and Excel spreadsheet: H-based methods. *Comput. Appl. Eng. Educ.* **2017**, *25*, 129–141. [CrossRef]
15. Epp, R.; Fowler, A.G. Efficient code for steady flows in networks. *J. Hydraul. Div. ASCE* **1970**, *96*, 43–56.
16. Praks, P.; Brkić, D. Choosing the optimal multi-point iterative method for the Colebrook flow friction equation. *Processes* **2018**, *6*, 130. [CrossRef]
17. Praks, P.; Brkić, D. Advanced iterative procedures for solving the implicit Colebrook equation for fluid flow friction. *Adv. Civ. Eng.* **2018**, *2018*, 5451034. [CrossRef]

18. Brkić, D. An improvement of Hardy Cross method applied on looped spatial natural gas distribution networks. *Appl. Energy* **2009**, *86*, 1290–1300. [CrossRef]
19. Coelho, P.M.; Pinho, C. Considerations about equations for steady state flow in natural gas pipelines. *J. Braz. Soc. Mech. Sci. Eng.* **2007**, *29*, 262–273. [CrossRef]
20. Bagajewicz, M.; Valtinson, G. Computation of natural gas pipeline hydraulics. *Ind. Eng. Chem. Res.* **2014**, *53*, 10707–10720. [CrossRef]
21. Pambour, K.A.; Cakir Erdener, B.; Bolado-Lavin, R.; Dijkema, G.P.J. Development of a simulation framework for analyzing security of supply in integrated gas and electric power systems. *Appl. Sci.* **2017**, *7*, 47. [CrossRef]
22. Praks, P.; Kopustinskas, V.; Masera, M. Probabilistic modelling of security of supply in gas networks and evaluation of new infrastructure. *Reliab. Eng. Syst. Saf.* **2015**, *144*, 254–264. [CrossRef]
23. Colebrook, C.F. Turbulent flow in pipes with particular reference to the transition region between the smooth and rough pipe laws. *J. Inst. Civ. Eng.* **1939**, *11*, 133–156. [CrossRef]
24. Brkić, D.; Praks, P. Unified friction formulation from laminar to fully rough turbulent flow. *Appl. Sci.* **2018**, *8*, 2036. [CrossRef]
25. Corfield, G.; Hunt, B.E.; Ott, R.J.; Binder, G.P.; Vandaveer, F.E. Distribution design for increased demand. In *Gas Engineers Handbook*; Segeler, C.G., Ed.; Industrial Press: New York, NY, USA, 1974; Chapter 9; pp. 63–83.
26. Brkić, D. Discussion of "Economics and statistical evaluations of using Microsoft Excel Solver in pipe network analysis" by Oke, I.A.; Ismail, A.; Lukman, S.; Ojo, S.O.; Adeosun, O.O.; Nwude, M.O. *J. Pipeline Syst. Eng.* **2018**, *9*, 07018002. [CrossRef]
27. Džunić, J.; Petković, M.S.; Petković, L.D. A family of optimal three-point methods for solving nonlinear equations using two parametric functions. *Appl. Math. Comput.* **2011**, *217*, 7612–7619. [CrossRef]
28. Khdhr, F.W.; Saeed, R.K.; Soleymani, F. Improving the Computational Efficiency of a Variant of Steffensen's Method for Nonlinear Equations. *Mathematics* **2019**, *7*, 306. [CrossRef]
29. Behl, R.; Salimi, M.; Ferrara, M.; Sharifi, S.; Alharbi, S.K. Some Real-Life Applications of a Newly Constructed Derivative Free Iterative Scheme. *Symmetry* **2019**, *11*, 239. [CrossRef]
30. Hamam, Y.M.; Brameller, A. Hybrid method for the solution of piping networks. *Proc. Inst. Electr. Eng.* **1971**, *118*, 1607–1612. [CrossRef]
31. Wood, D.J.; Charles, C.O.A. Hydraulic network analysis using linear theory. *J. Hydraul. Div. ASCE* **1972**, *98*, 1157–1170.
32. Wood, D.J.; Rayes, A.G. Reliability of algorithms for pipe network analysis. *J. Hydraul. Div. ASCE* **1981**, *107*, 1145–1161.
33. Brkić, D.; Praks, P. An efficient iterative method for looped pipe network hydraulics free of flow-corrections. *Fluids* **2019**, *4*, 73. [CrossRef]
34. Shamir, U.; Howard, C.D.D. Water distribution systems analysis. *J. Hydraul. Div. ASCE* **1968**, *94*, 219–234.
35. Lopes, A.M.G. Implementation of the Hardy-Cross method for the solution of piping networks. *Comput. Appl. Eng. Educ.* **2004**, *12*, 117–125. [CrossRef]
36. Elaoud, S.; Hafsi, Z.; Hadj-Taieb, L. Numerical modelling of hydrogen-natural gas mixtures flows in looped networks. *J. Pet. Sci. Eng.* **2017**, *159*, 532–541. [CrossRef]
37. Bermúdez, J.-R.; López-Estrada, F.-R.; Besançon, G.; Valencia-Palomo, G.; Torres, L.; Hernández, H.-R. Modeling and simulation of a hydraulic network for leak diagnosis. *Math. Comput. Appl.* **2018**, *23*, 70. [CrossRef]
38. Adedeji, K.B.; Hamam, Y.; Abe, B.T.; Abu-Mahfouz, A.M. Leakage detection and estimation algorithm for loss reduction in water piping networks. *Water* **2017**, *9*, 773. [CrossRef]
39. Todini, E.; Rossman, L.A. Unified framework for deriving simultaneous equation algorithms for water distribution networks. *J. Hydraul. Eng.* **2012**, *139*, 511–526. [CrossRef]
40. Ghazanfari, S.A.; Wahid, M.A. Heat transfer enhancement and pressure drop for fin-and-tube compact heat exchangers with delta winglet-type vortex generators. *Facta Univ. Ser. Mech. Eng.* **2018**, *16*, 233–247. [CrossRef]
41. Brkić, D. Review of explicit approximations to the Colebrook relation for flow friction. *J. Pet. Sci. Eng.* **2011**, *77*, 34–48. [CrossRef]
42. Brkić, D. Lambert W function in hydraulic problems. *Math. Balk.* **2012**, *26*, 285–292. Available online: http://www.math.bas.bg/infres/MathBalk/MB-26/MB-26-285-292.pdf (accessed on 29 March 2019).

43. Brkić, D.; Ćojbašić, Ž. Evolutionary optimization of Colebrook's turbulent flow friction approximations. *Fluids* **2017**, *2*, 15. [CrossRef]
44. Praks, P.; Brkić, D. One-log call iterative solution of the Colebrook equation for flow friction based on Padé polynomials. *Energies* **2018**, *11*, 1825. [CrossRef]
45. Brkić, D.; Praks, P. Accurate and Efficient Explicit Approximations of the Colebrook Flow Friction Equation Based on the Wright ω-Function. *Mathematics* **2019**, *7*, 34. [CrossRef]
46. Brkić, D. Solution of the implicit Colebrook equation for flow friction using Excel. *Spreadsheets Educ.* **2017**, *10*, 4663. Available online: https://sie.scholasticahq.com/article/4663.pdf (accessed on 29 March 2019).
47. Olivares Gallardo, A.P.; Guerra Rojas, R.A.; Alfaro Guerra, M.A. Evaluación experimental de la solución analítica exacta de la ecuación de Colebrook-White. *Ing. Investig. Y Tecnol.* **2019**, *20*, 1–11. (In Spanish) [CrossRef]
48. Eaton, L.K. *Hardy Cross: American Engineer*; University of Illinois Press: Champaign, IL, USA, 2006.
49. Eaton, L.K. Hardy Cross and the "Moment Distribution Method". *Nexus Netw. J.* **2001**, *3*, 15–24. [CrossRef]
50. Weingardt, R.G. Hardy Cross, A man ahead of his time. *Struct. Mag.* **2005**, *3*, 40–41.
51. Weingardt, R.G. Hardy Cross and Albert A. Dorman. *Leadersh. Manag. Eng.* **2004**, *4*, 51–54. [CrossRef]
52. Cross, H. *Engineers and Ivory Towers*; McGraw-Hill: New York, NY, USA, 1952.
53. Volokh, K.Y. On foundations of the Hardy Cross method. *Int. J. Solids Struct.* **2002**, *39*, 4197–4200. [CrossRef]
54. Baugh, J.; Liu, S. A general characterization of the Hardy Cross method as sequential and multiprocess algorithms. *Structures* **2016**, *6*, 170–181. [CrossRef]
55. Zweig, C.M. Hardy Cross, The Tinkerer. *Civ. + Struct. Eng. Mag.* **2014**. Available online: https://csengineermag.com/article/hardy-cross-the-tinkerer/ (accessed on 10 April 2019).
56. Cross, H. Analysis of continuous frames by distributing fixed-end moments. *Am. Soc. Civ. Eng. Trans.* **1932**, *96*, 1793.
57. Fresl, K.; Gidak, P.; Hak, S. Iz povijesti razvoja iteracijskih postupaka (From the history of development of iterative procedures). *Građevinar* **2010**, *62*, 959–970. Available online: https://hrcak.srce.hr/61473 (accessed on 10 April 2019). (In Serbian).
58. Čališev, K. Izračunavanje višestruko statički neodređenih sistema pomoću postepenih aproksimacija. *Teh. List Udruženja Jugosl. Inženjera Arhit.* **1923**, *5*, 17–21. (In Serbian)
59. Timošenko, S. *History of Strength of Materials: With a Brief Account of the History of Theory of Elasticity and Theory of Structures*; McGraw-Hill: New York, NY, USA, 1953; ISSN 0486611876.
60. Samuelsson, A.; Zienkiewicz, O.C. History of the stiffness method. *Int. J. Numer. Methods Eng.* **2006**, *67*, 149–157. [CrossRef]

applied sciences

MDPI

Article

Multi-Switching Combination Synchronization of Three Fractional-Order Delayed Systems

Bo Li [1], Yun Wang [1,*] and Xiaobing Zhou [2]

[1] School of Computer Science and Engineering, Southeast University, Nanjing 211189, China; boli@seu.edu.cn
[2] School of Information Science and Engineering, Yunnan University, Kunming 650091, China; zhouxb@ynu.edu.cn
* Correspondence: ywang_cse@seu.edu.cn

Received: 19 July 2019; Accepted: 13 August 2019; Published: 15 October 2019

Abstract: Multi-switching combination synchronization of three fractional-order delayed systems is investigated. This is a generalization of previous multi-switching combination synchronization of fractional-order systems by introducing time-delays. Based on the stability theory of linear fractional-order systems with multiple time-delays, we propose appropriate controllers to obtain multi-switching combination synchronization of three non-identical fractional-order delayed systems. In addition, the results of our numerical simulations show that they are in accordance with the theoretical analysis.

Keywords: multi-switching combination synchronization; time-delay; fractional-order; stability

1. Introduction

Fractional calculus has attracted researchers from various fields due to fractional dimensions widely existing in nature and engineering fields [1–3]. Compared to the integer-order dynamical systems, the fractional-order counterparts can exhibit more complex dynamical behaviors. Some of the researches on integer-order dynamical systems can be generalized to fractional-order dynamical systems. Fractional-order dynamical systems have been widely investigated, such as synchronization [4], identification [5], stabilization [6] and approximate entropy analysis [7,8]. Time-delay is a frequently encountered phenomenon in real applications, such as physical, communication, economical, pneumatic and biological systems [9]. Introducing time-delay into a system can enrich its dynamic characteristics and describe a real-life phenomenon more precisely. Thus, the fractional-order delayed differential equation (FDDE) is becoming a hot topic for scientists and engineers, and it has many theoretical and practical applications [10]. Nowadays, the chaotic behavior and synchronization of FDDE attract intensive research interests. In [11], Bhalekar et al. introduced the fractional-order delayed Liu system. The fractional-order delayed financial system was presented in [12], and hybrid projective synchronization between the aforementioned two systems was achieved in [13]. The fractional-order delayed Chen system was proposed in [14], while its adaptive synchronization was investigated in [15]. The fractional-order delayed porous media was proposed in [16]. In [17], a fractional-order delayed Newton–Leipnik system was taken as an example to present intermittent synchronizing delayed fractional nonlinear system.

Due to its wide applications in secure communication, synchronization of fractional-order delayed chaotic systems are extensively investigated [18–21]. However, all the above-mentioned and other synchronization schemes are in traditional drive–response ways, which only have unique drive and response systems. Recently, Luo et al. [22] extended the traditional drive–response synchronization to combination synchronization, which has two drive systems and one response system. Compared to the drive–response synchronization, combination synchronization has stronger anti-decode and anti-attack abilities in image encryption and secure communication, in which origin messages are split into two parts and each part can be embedded into two separate drive systems. There are many works on combination synchronization [23–26]. To further strengthen the security in secure communication, Vincent et al. [27] proposed multi-switching combination synchronization scheme, in which the two drive systems are synchronized with the response system in different states. Based on the nonlinear control technique, Zheng [28] studied multi-switching combination synchronization of three non-identical chaotic systems. Khan [29] investigated adaptive multi-switching combination synchronization among three non-identical chaotic systems. In [30], Ahmad et al. proposed globally exponential multi-switching combination synchronization scheme, and applied it to secure communications. Multi-switching combination synchronization was applied in encrypted audio communication in [31]. The previous work on multi-switching combination synchronization schemes are based on integral-order chaotic systems. To improve the security in these synchronization schemes, based on fractional-order chaotic systems, Bhat et al. [32] extended the work in [27] to study multi-switching combination synchronization among three non-identical fractional-order chaotic systems. Khan et al. [33] investigated multi-switching combination-combination synchronization among a class of four non-identical fractional-order chaotic systems. In multi-switching combination-combination synchronization scheme, the state variables of two drive systems synchronize with different state variables of two response systems simultaneously, which makes the security of this scheme higher than that in [32].

Since time-delay is a frequently encountered phenomenon in real applications, and the time-delay can be used as an additional parameter in synchronization to increase security in secure communication, we consider the work in [32] to investigate multi-switching combination synchronization scheme for non-identical fractional-order delayed systems by introducing time-delays in fractional-order systems.

The rest of this paper is organized as follows. In Section 2, the concept of fractional calculus and the stability theory of linear fractional-order systems with multiple time-delays are briefly introduced. The multi-switching combination synchronization scheme of three non-identical fractional-order delayed systems is analyzed in Section 3. In Section 4, numerical simulations performed using MATLAB are presented. Finally, conclusions are drawn in Section 5.

2. Preliminaries

Fractional calculus is a generalization of integration and differentiation to non-integer order fundamental operator ${}_aD_t^r$, which is defined as

$$ {}_aD_t^r = \begin{cases} \frac{d^r}{dt^r} & : r > 0, \\ 1 & : r = 0, \\ \int_a^t (d\tau)^{-r} & : r < 0. \end{cases} \tag{1} $$

There are several different definitions for the fractional-order differential operator [34]. Because the Caputo definition is easy to understand and is frequently used in the literature, we apply this definition in this paper, which is

$$ {}_aD_t^r f(t) = \frac{1}{\Gamma(n-r)} \int_a^t \frac{f(\tau)}{t-\tau}^{r-n+1} d\tau \tag{2} $$

where $1 < r < n$.

Lemma 1 ([35]). *Suppose $f(t) \in C_a^\alpha([a,b]), D_a^\alpha f(t) \in C_a^\beta([a,b]), \alpha > 0, \beta > 0, m-1 < \beta < m, n-1 < \alpha < n$, then*

$$ {}_a^CD_t^\beta \left({}_a^CD_t^\alpha f(t) \right) = {}_a^C D_t^\alpha ({}_a^CD_t^\beta f(t)) = {}_a^C D_t^{\alpha+\beta} f(t) \tag{3} $$

Lemma 2 ([35]). *Consider*

$$ {}_{a}^{C}D_{t}^{\alpha}x^{n} = \frac{\Gamma(n+1)x^{n-\alpha}}{\Gamma(n+1-\alpha)} {}_{a}^{C}D_{t}^{\alpha}x \tag{4} $$

Lemma 3 ([35]). *Let $\phi(t) \in R$ be a continuous and derivable function. Then, for any time instant $t \geq t_0$*

$$ \frac{1}{2}D^{q}\phi^{2}(t) = \phi(t)D^{q}\phi(t),\ \forall q \in (0,1) \tag{5} $$

Given the following n-dimensional linear fractional-order system with multiple time-delays [36]:

$$ \begin{cases} D^{q_1}y_1(t) = \alpha_{11}y_1(t-\tau_{11}) + \alpha_{12}y_2(t-\tau_{12}) \\ \qquad + \cdots + \alpha_{1n}y_n(t-\tau_{1n}), \\ D^{q_2}y_2(t) = \alpha_{21}y_1(t-\tau_{21}) + \alpha_{22}y_2(t-\tau_{22}) \\ \qquad + \cdots + \alpha_{2n}y_n(t-\tau_{2n}), \\ \vdots \\ D^{q_n}y_n(t) = \alpha_{n1}y_1(t-\tau_{n1}) + \alpha_{n2}y_2(t-\tau_{n2}) \\ \qquad + \cdots + \alpha_{nn}y_n(t-\tau_{nn}), \end{cases} \tag{6} $$

where $q_i \in (0,1)$ is the fractional-derivative order, $y_i(t)$ is the state, and $\tau_{ij} > 0$ is the time-delay, the initial value $y_i(t) = \phi_i(t)$ is given by $-max\tau_{ij} = -\tau_{max} \leq t \leq 0$, $A = [a_{ij}] \in R_{n\times n}$ is the coefficient matrix.

Performing Laplace transform on the system in Equation (6) yields

$$ \Delta(s) \cdot Y(s) = b(s), \tag{7} $$

where $Y(s) = (Y_1(s),\ Y_2(s),\ ...,\ Y_n(s))^T$ is the Laplace transform of $y(t) = (y_1(t),\ y_2(t),\ ...,\ y_n(t))^T$, $b(s) = (b_1(s),\ b_2(s),\ ...,\ b_n(s))^T$ is the remaining non-linear part, the characteristic matrix of the system in Equation (6) is

$$ \Delta(s) = \begin{pmatrix} s^{q_1} - \alpha_{11}e^{-s\tau_{11}} & -\alpha_{12}e^{-s\tau_{12}} & \cdots & -\alpha_{1n}e^{-s\tau_{1n}} \\ -\alpha_{21}e^{-s\tau_{21}} & s^{q_2} - \alpha_{22}e^{-s\tau_{22}} & \cdots & -\alpha_{2n}e^{-s\tau_{2n}} \\ \vdots & \vdots & \ddots & \vdots \\ -\alpha_{n1}e^{-s\tau_{n1}} & -\alpha_{n2}e^{-s\tau_{n2}} & \cdots & s^{q_n} - \alpha_{nn}e^{-s\tau_{nn}} \end{pmatrix}. \tag{8} $$

Theorem 1 ([36]). *If all the roots of the characteristic equation $\det(\Delta(s)) = 0$ have negative real parts, then the zero solution of the system in Equation (6) is Lyapunov globally asymptotically stable.*

Corollary 1 ([36]). *If $q_1 = q_2 = \cdots = q_n = \beta \in (0,1)$, all the eigenvalues λ of the coefficient matrix A satisfy $|\arg(\lambda)| > \beta\pi/2$, and the characteristic equation $\det(\Delta(s)) = 0$ has no purely imaginary roots for any $\tau_{ij} > 0,\ i,j = 1,2,\ldots,n$, then the zero solution of the system in Equation (6) is Lyapunov globally asymptotically stable.*

3. Multi-Switching Combination Synchronization Scheme

Multi-switching combination synchronization among three non-identical fractional-order delayed systems is investigated in this section.

The two drive systems are

$$ \begin{aligned} D^{\alpha}x(t) &= x(t) + x(t-\tau) + A(x(t), x(t-\tau)), \\ x(t) &= x(0), t \in [-\tau, 0], \end{aligned} \tag{9} $$

and

$$
\begin{aligned}
D^{\alpha}y(t) &= y(t) + y(t-\tau) + B(y(t), y(t-\tau)),\\
y(t) &= y(0), t \in [-\tau, 0].
\end{aligned}
\tag{10}
$$

The response system is

$$
\begin{aligned}
D^{\alpha}z(t) &= z(t) + z(t-\tau) + C(z(t), z(t-\tau)) + U,\\
z(t) &= z(0), t \in [-\tau, 0],
\end{aligned}
\tag{11}
$$

in which, $\alpha \in (0,1)$ is the fractional order, $\tau > 0$ is the time-delay, $U = (U_1, ..., U_n)$ is the controller vector, $x = (x_1, ..., x_n)^T \in R^n$, $y = (y_1, ..., y_n)^T \in R^n$ and $z = (z_1, ..., z_n)^T \in R^n$ are state vectors, and $A : R^{2n} \to R^n$, $B : R^{2n} \to R^n$ and $C : R^{2n} \to R^n$ are continuous vector functions.

Define the error state as $e_{klm} = f_k z_k - g_l x_l - h_m y_m (k, l, m = 1, ..., n)$. Then, we have the error state vector

$$
e(t) = Fz - Gx - Hy, \tag{12}
$$

where $e(t)$ is the vector form of e_{klm}, $F = diag\{f_1, f_2, \ldots, f_n\} \in R^{n\times n}$, $G = diag\{g_1, g_2, \ldots, g_n\} \in R^{n\times n}$ and $H = diag\{h_1, h_2, \ldots, h_n\} \in R^{n\times n}$ are real scaling matrix. Accordingly, $e_{klm}(t-\tau) = f_k z_k(t-\tau) - g_l x_l(t-\tau) - h_m y_m(t-\tau)$.

Definition 1 ([27]). *The systems in Equations (9) and (10) and the system in Equation (11) are defined to be multi-switching combination synchronization if F, G, H are non-zeros, and $k \neq l \neq m$, $k = l \neq m$, $k \neq l = m$, $k = m \neq l$, such that:*

$$
\lim_{t\to+\infty} \| e(t) \| = \lim_{t\to+\infty} \| Fz - Gx - Hy \| = 0 \tag{13}
$$

where $\| . \|$ represents the matrix norm.

Remark 1. *If $k = l = m$, the systems in Equations (9) and (10) and the system in Equation (11) are defined to be combination synchronization [22].*

Remark 2. *If the scaling matrix $F \neq 0$, $G = 0$ or $H = 0$, the multi-switching combination synchronization mentioned above is simplified to multi-switching hybrid projective synchronization.*

From the systems in Equations (9)–(11), we have the error system as follows

$$
D^{\alpha}e(t) = FD^{\alpha}z(t) - GD^{\alpha}x(t) - HD^{\alpha}y(t) \tag{14}
$$

To achieve multi-switching combination synchronization among the above systems, a non-linear controller is constructed:

$$
U = \tilde{K}e(t) + GA(x(t), x(t-\tau)) + HB(y(t), y(t-\tau)) - FC(z(t), z(t-\tau)), \tag{15}
$$

where $\tilde{K} = K - I$, I is an n-dimensional unit matrix, $K = diag\{k_1, k_2, \ldots, k_n\}$ is a feedback gain matrix.

Substituting the systems in Equations (9)–(11) and (15) into the system in Equation (14), we have

$$
D^{\alpha}e(t) = (\tilde{k} + I)e(t) + e(t-\tau) = Ke(t) + e(t-\tau). \tag{16}
$$

Thus, the multi-switching combination synchronization between the systems in Equations (9) and (10) and the system in Equation (11) is changed into the analysis of the asymptotical stability of the system in Equation (16).

In light of Corollary 1, a sufficient condition to achieve multi-switching combination synchronization between the systems in Equations (9) and (10) and the system in Equation (11) is obtained as follows.

Proposition 1. *Multi-switching combination synchronization between the systems in Equations (9) and (10) and the system in Equation (11) can be achieved if there exists a matrix $K = diag\{k_1, k_2, \ldots, k_n\}$ in the system in Equation (16) such that $k_i < -1/\sin(\alpha\pi/2)$, $(i = 1, 2, \ldots, n)$.*

Proof. For the system in Equation (16), $A = K + I$ is the coefficient matrix. Since $k_i < -1/\sin(\alpha\pi/2), \alpha \in (0,1)$, the eigenvalues of A are $\lambda_i = k_i + 1 < 0$, $(i = 1, 2, \ldots, n)$. Then, $|\arg(\lambda)| > \pi/2 > \alpha\pi/2$ holds.

Performing Laplace transform on the system in Equation (16) yields

$$\Delta(s) \cdot E(s) = s^{\alpha-1}e(0) + e(0)e^{-s\tau}\int_{-\tau}^{0} e^{-s\tau}\, dx, \tag{17}$$

where $E(s)$ is the Laplace transform of $e(t)$, $e(0) = Fz(0) - Gx(0) - Hy(0)$, $\Delta(s) = s^{a}I - K - e^{-s\tau}I$ is the characteristic matrix. Then,

$$det(\Delta(s)) = |s^{\alpha}I - K - e^{-s\tau}I| = (s^{a} - k_1 - e^{-s\tau})(s^{a} - k_2 - e^{-s\tau})\ldots(s^{a} - k_n - e^{-s\tau}) = 0. \tag{18}$$

Assume

$$(s^{a} - k_i - e^{-s\tau}) = 0, i = 1, 2, \ldots, n. \tag{19}$$

has a root $s = wi = |w|\,(\cos(\pi/2) + i\sin(\pm\pi/2))$. Thus,

$$|w|^{\alpha}\,(\cos(\alpha\pi/2) + i\sin(\pm\alpha\pi/2)) - k_i - \cos(\omega\tau) + i\sin(\omega\tau) = 0. \tag{20}$$

From the above equation, we can get

$$\begin{aligned} |w|^{\alpha}\cos(\alpha\pi/2) - k_i &= \cos(\omega\tau), \\ |w|^{\alpha}\sin(\pm\alpha\pi/2) &= -\sin(\omega\tau). \end{aligned} \tag{21}$$

Hence,

$$|w|^{2\alpha} - 2k_i\cos(\alpha\pi/2)\,|w|^{\alpha} + k_i^2 - 1 = 0. \tag{22}$$

Since $k_i < -1/\sin(\alpha\pi/2), \alpha \in (0,1)$, the discriminant of the roots satisfies

$$\begin{aligned} \Delta &= (-2k_i\cos(\alpha\pi/2))^2 - 4(k_i^2 - 1) \\ &= 4(1 - k_i^2\sin^2(\alpha\pi/2)) \\ &< 0. \end{aligned} \tag{23}$$

Then, Equation (22) has no real solutions, and Equation (18) has no purely imaginary roots.

In light of Corollary 1, the zero solution of the system in Equation (16) is globally asymptotically stable, i.e., multi-switching combination synchronization is obtained between the systems in Equations (9) and (10) and the system in Equation (11). □

4. Numerical Examples

Numerical simulations were carried out to illustrate the above proposed multi-switching combination synchronization scheme. We used the same systems as in [32] with time-delays, which are fractional-order delayed Lorenz, Chen, and Rössler systems, and the numerical simulations were carried out in MATLAB.

The fractional-order delayed Lorenz system [37] was considered as the first drive system

$$\begin{cases} D^\alpha x_1 = a_1(x_2 - x_1), \\ D^\alpha x_2 = c_1 x_1 - x_2 - x_1 x_3, \\ D^\alpha x_3 = x_1 x_2 - b_1 x_3(t-\tau). \end{cases} \tag{24}$$

The system in Equation (24) exhibits a chaotic attractor, as illustrated in Figure 1. The system in Equation (24) can be rewritten as

$$\begin{aligned} D^\alpha x(t) &= x(t) + x(t-\tau) + A(x(t), x(t-\tau)), \\ x(t) &= x(0), t \in [-\tau, 0], \end{aligned} \tag{25}$$

where

$$A(x(t), x(t-\tau)) = \begin{pmatrix} a_1 x_2 - (a_1+1)x_1 - x_1(t-\tau) \\ c_1 x_1 - 2x_2 - x_1 x_3 - x_2(t-\tau) \\ x_1 x_2 - (b_1+1)x_3(t-\tau) - x_3 \end{pmatrix}. \tag{26}$$

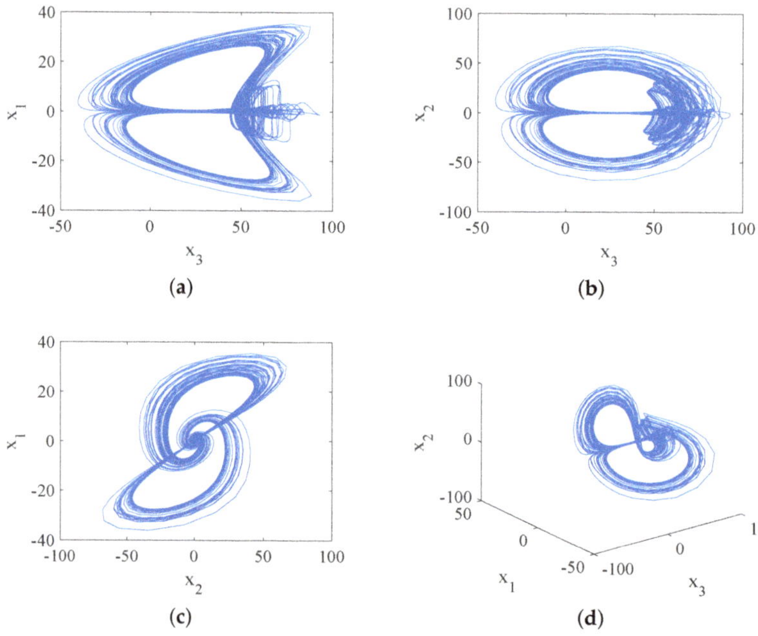

Figure 1. Chaotic attractor of Lorenz system with $\alpha = 0.95$, $\tau = 0.4$: (**a**) $x_3 - x_1$ plane; (**b**) $x_3 - x_2$ plane; (**c**) $x_2 - x_1$ plane; and (**d**) $x_3 - x_1 - x_2$ space.

The fractional-order delayed Chen system [14] is the second drive system

$$\begin{cases} D^\alpha y_1 = a_2(y_2 - y_1), \\ D^\alpha y_2 = (c_2 - a_2)y_1 - y_1 y_3 + c_2 y_2, \\ D^\alpha y_3 = y_1 y_2 - b_2 y_3(t-\tau). \end{cases} \tag{27}$$

The system in Equation (27) displays a chaotic attractor, as shown in Figure 2. We rewrite the system in Equation (27) as

$$\begin{aligned} D^{\alpha}y(t) &= y(t) + y(t-\tau) + B(y(t), y(t-\tau)), \\ y(t) &= y(0), t \in [-\tau, 0], \end{aligned} \tag{28}$$

where

$$B(y(t), y(t-\tau)) = \begin{pmatrix} a_2(y_2 - y_1) - y_1 - y_1(t-\tau) \\ (c_2 - a_2)y_1 - y_1y_3 + (c_2 - 1)y_2 - y_2(t-\tau) \\ y_1y_2 - (b_2 + 1)y_3(t-\tau) - y_3 \end{pmatrix}. \tag{29}$$

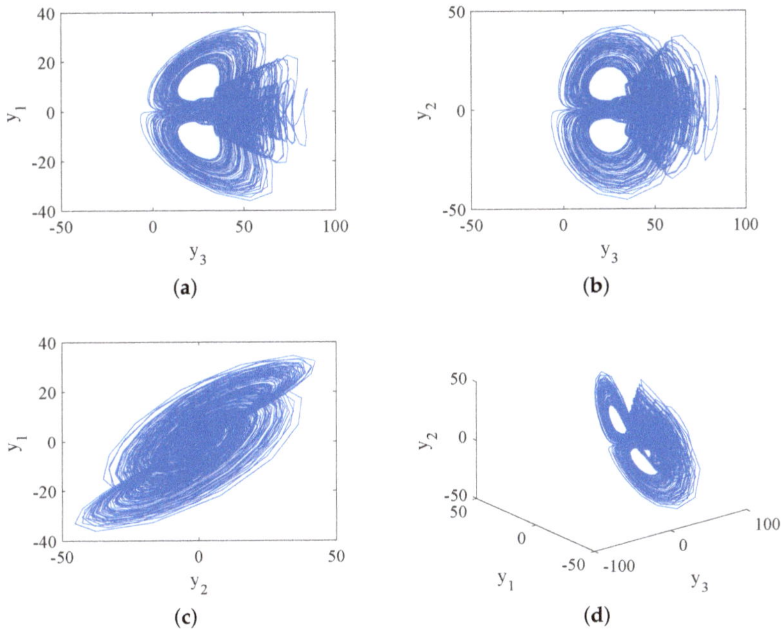

Figure 2. Chaotic attractor of Chen system with $\alpha = 0.95$, $\tau = 0.4$: (**a**) $y_3 - y_1$ plane; (**b**) $y_3 - y_2$ plane; (**c**) $y_2 - y_1$ plane; and (**d**) $y_3 - y_1 - y_2$ space.

The fractional-order delayed Rössler system is the response system, given by

$$\begin{cases} D^{\alpha}z_1 = -(z_2 + z_3) + 0.2z_1(t-\tau) + U_1, \\ D^{\alpha}z_2 = z_1 + a_3z_2 + U_2, \\ D^{\alpha}z_3 = z_3(z_1 - m_3) + b_3 + U_3. \end{cases} \tag{30}$$

where U_1, U_2 and U_3 are determined afterwards. Without the controllers, the system in Equation (30) exhibits a chaotic attractor, as illustrated in Figure 3. The system in Equation (30) is rewritten as

$$\begin{aligned} D^{\alpha}z(t) &= z(t) + z(t-\tau) + C(z(t), z(t-\tau)) + U, \\ z(t) &= z(0), t \in [-\tau, 0], \end{aligned} \tag{31}$$

where

$$C(z(t), z(t-\tau)) = \begin{pmatrix} -(z_2+z_3) - 0.8z_1(t-\tau) - z_1 \\ z_1 + (a_3-1)z_2 - z_2(t-\tau) \\ z_3(z_1-m_3) + b_3 - z_3 - z_3(t-\tau) \end{pmatrix}. \tag{32}$$

For the systems in Equations (24), (27) and (30), there are eight possible switch combination synchronization cases.

When $k \neq l \neq m$, we have e_{123}, e_{231}, e_{312} and e_{132}, e_{213}, e_{321}.

When $k = m \neq l$, we have e_{121}, e_{232}, e_{313} and e_{131}, e_{212}, e_{323}.

When $k \neq l = m$, we have e_{122}, e_{233}, e_{311} and e_{133}, e_{211}, e_{322}.

When $k = l \neq m$, we have e_{112}, e_{223}, e_{331} and e_{113}, e_{221}, e_{332}.

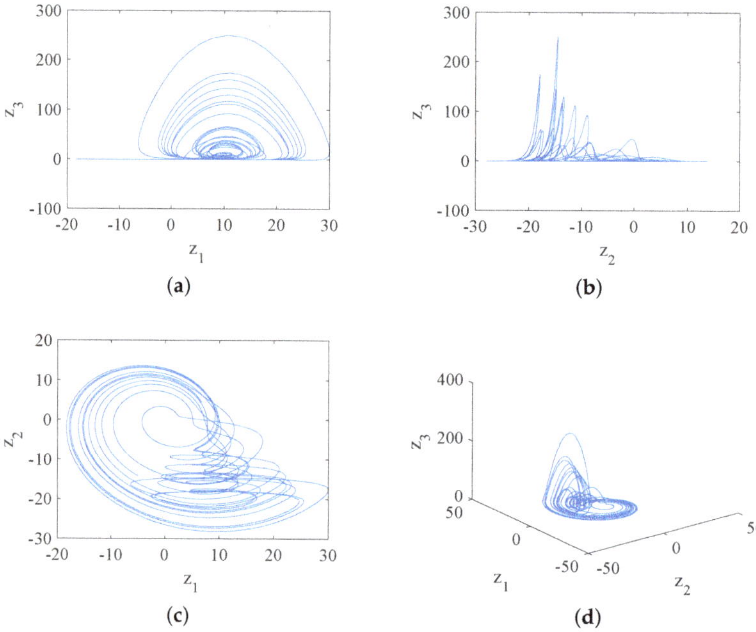

Figure 3. Chaotic attractor of Rössler system with $\alpha = 0.95$, $\tau = 0.4$: (**a**) $z_1 - z_3$ plane; (**b**) $z_2 - z_3$ plane; (**c**) $z_1 - z_2$ plane; and (**d**) $z_2 - z_1 - z_3$ space.

We randomly pick two cases

$$\begin{cases} e_{123} = f_1z_1 - g_2x_2 - h_3y_3, \\ e_{231} = f_2z_2 - g_3x_3 - h_1y_1, \quad \textbf{case 1} \\ e_{312} = f_3z_3 - g_1x_1 - h_2y_2. \end{cases} \tag{33}$$

and

$$\begin{cases} e_{112} = f_1z_1 - g_1x_1 - h_2y_2, \\ e_{223} = f_2z_2 - g_2x_2 - h_3y_3, \quad \textbf{case 2} \\ e_{331} = f_3z_3 - g_3x_3 - h_1y_1. \end{cases} \tag{34}$$

In the following, we analyze these two cases in detail.

Case 1

From the systems in Equations (24), (27) and (30), we have the error dynamical system

$$\begin{cases} D^{\alpha}e_{123} = f_1 D^{\alpha}z_1 - g_2 D^{\alpha}x_2 - h_3 D^{\alpha}y_3, \\ D^{\alpha}e_{231} = f_2 D^{\alpha}z_2 - g_3 D^{\alpha}x_3 - h_1 D^{\alpha}y_1, \\ D^{\alpha}e_{312} = f_3 D^{\alpha}z_3 - g_1 D^{\alpha}x_1 - h_2 D^{\alpha}y_2. \end{cases} \tag{35}$$

such that

$$\begin{cases} \lim\limits_{t\to+\infty} \| f_1z_1 - g_2x_2 - h_3y_3 \| = 0, \\ \lim\limits_{t\to+\infty} \| f_2z_2 - g_3x_3 - h_1y_1 \| = 0, \\ \lim\limits_{t\to+\infty} \| f_3z_3 - g_1x_1 - h_2y_2 \| = 0. \end{cases} \tag{36}$$

Substituting the systems in Equations (24), (27) and (30) into the system in Equation (35) yields

$$\begin{cases} D^{\alpha}e_{123} = f_1(-(z_2+z_3) + 0.2z_1(t-\tau) + U_1) - g_2(c_1x_1 - x_2 - x_1x_3) \\ \qquad - h_3(y_1y_2 - b_2y_3(t-\tau)), \\ D^{\alpha}e_{231} = f_2(z_1 + a_3z_2 + U_2) - g_3(x_1x_2 - b_1x_3(t-\tau)) \\ \qquad - h_1(a_2(y_2-y_1)), \\ D^{\alpha}e_{312} = f_3(z_3(z_1 - m_3) + b_3 + U_3) - g_1(a_1(x_2-x_1)) \\ \qquad - h_2((c_2-a_2)y_1 - y_1y_3 + c_2y_2). \end{cases} \tag{37}$$

Here, we obtain the following results.

Theorem 2. *Multi-switching combination synchronization between the systems in Equations (24) and (27) and the system in Equation (30) can be achieved with the following controllers*

$$\begin{cases} U_1 = \dfrac{1}{f_1}\{(k_1-1)(f_1z_1 - g_2x_2 - h_3y_3) + g_2(c_1x_1 - 2x_2 - x_1x_3 - x_2(t-\tau)) \\ \quad + h_3(y_1y_2 - (b_2+1)y_3(t-\tau) - y_3) - f_1(-(z_2+z_3) - 0.8z_1(t-\tau) - z_1)\}, \\ U_2 = \dfrac{1}{f_2}\{(k_2-1)(f_2z_2 - g_3x_3 - h_1y_1) + g_3(x_1x_2 - (b_1+1)x_3(t-\tau) - x_3) \\ \quad + h_1(a_2(y_2-y_1) - y_1 - y_1(t-\tau)) - f_2(z_1 + (a_3-1)z_2 - z_2(t-\tau))\}, \\ U_3 = \dfrac{1}{f_3}\{(k_3-1)(f_3z_3 - g_1x_1 - h_2y_2) + g_1(a_1x_2 - (a_1+1)x_1 - x_1(t-\tau)) \\ \quad + h_2((c_2-a_2)y_1 - y_1y_3 + (c_2-1)y_2 - y_2(t-\tau)) \\ \quad - f_3(z_3(z_1 - m_3) + b_3 - z_3 - z_3(t-\tau))\}. \end{cases} \tag{38}$$

Supposing $F \neq 0$ and $G = 0$ or $H = 0$, we have the following results.

Corollary 2. *(i) Supposing that $f_i \neq 0$, $g_i = 0$ and $h_i \neq 0$ $(i = 1,2,3)$, multi-switching hybrid projective synchronization between the systems in Equations (27) and (30) can be achieved with the following controllers*

$$\begin{cases} U_1 = \frac{1}{f_1}\{(k_1-1)(f_1z_1 - h_3y_3) + h_3(y_1y_2 - (b_2+1)y_3(t-\tau) - y_3) \\ \quad - f_1(-(z_2+z_3) - 0.8z_1(t-\tau) - z_1)\}, \\ U_2 = \frac{1}{f_2}\{(k_2-1)(f_2z_2 - h_1y_1) + h_1(a_2(y_2-y_1) - y_1 - y_1(t-\tau)) \\ \quad - f_2(z_1 + (a_3-1)z_2 - z_2(t-\tau))\}, \\ U_3 = \frac{1}{f_3}\{(k_3-1)(f_3z_3 - h_2y_2) + h_2((c_2-a_2)y_1 - y_1y_3 + (c_2-1)y_2 - y_2(t-\tau)) \\ \quad - f_3(z_3(z_1-m_3) + b_3 - z_3 - z_3(t-\tau))\}. \end{cases} \tag{39}$$

(ii) Similarly, supposing that $f_i \neq 0$, $g_i \neq 0$ and $h_i = 0$ $(i = 1,2,3)$, multi-switching hybrid projective synchronization between the systems in Equations (24) and (30) can be achieved with the following controllers

$$\begin{cases} U_1 = \frac{1}{f_1}\{(k_1-1)(f_1z_1 - g_2x_2) + g_2(c_1x_1 - 2x_2 - x_1x_3 - x_2(t-\tau)) \\ \quad - f_1(-(z_2+z_3) - 0.8z_1(t-\tau) - z_1)\}, \\ U_2 = \frac{1}{f_2}\{(k_2-1)(f_2z_2 - g_3x_3) + g_3(x_1x_2 - (b_1+1)x_3(t-\tau) - x_3) \\ \quad - f_2(z_1 + (a_3-1)z_2 - z_2(t-\tau))\}, \\ U_3 = \frac{1}{f_3}\{(k_3-1)(f_3z_3 - g_1x_1) + g_1(a_1x_2 - (a_1+1)x_1 - x_1(t-\tau)) \\ \quad - f_3(z_3(z_1-m_3) + b_3 - z_3 - z_3(t-\tau))\}. \end{cases} \tag{40}$$

Corollary 3. *Supposing that $f_i \neq 0$, $g_i = 0$ and $h_i = 0$ $(i = 1,2,3)$, the system in Equation (30) can be stabilized to its equilibrium $O(0,0,0)$ with the following controllers*

$$\begin{cases} U_1 = \frac{1}{f_1}\{(k_1-1)(f_1z_1) - f_1(-(z_2+z_3) - 0.8z_1(t-\tau) - z_1)\}, \\ U_2 = \frac{1}{f_2}\{(k_2-1)(f_2z_2) - f_2(z_1 + (a_3-1)z_2 - z_2(t-\tau))\}, \\ U_3 = \frac{1}{f_3}\{(k_3-1)(f_3z_3) - f_3(z_3(z_1-m_3) + b_3 - z_3 - z_3(t-\tau))\}. \end{cases} \tag{41}$$

Case 2

From the systems in Equations (24), (27) and (30), we have

$$\begin{cases} D^\alpha e_{112} = f_1D^\alpha z_1 - g_1D^\alpha x_1 - h_2D^\alpha y_2, \\ D^\alpha e_{223} = f_2D^\alpha z_2 - g_2D^\alpha x_2 - h_3D^\alpha y_3, \\ D^\alpha e_{331} = f_3D^\alpha z_3 - g_3D^\alpha x_3 - h_1D^\alpha y_1. \end{cases} \tag{42}$$

such that

$$\begin{cases} \lim_{t\to+\infty} \| f_1z_1 - g_1x_1 - h_2y_2 \| = 0, \\ \lim_{t\to+\infty} \| f_2z_2 - g_2x_2 - h_3y_3 \| = 0, \\ \lim_{t\to+\infty} \| f_3z_3 - g_3x_3 - h_1y_1 \| = 0. \end{cases} \tag{43}$$

Substituting the systems in Equations (24), (27), and (30) into the system in Equation (42) yields:

$$\begin{cases} D^{\alpha}e_{112} = f_1(-(z_2+z_3)+0.2z_1(t-\tau)+U_1) - g_1(a_1(x_2-x_1)) \\ \qquad - h_2((c_2-a_2)y_1 - y_1y_3 + c_2y_2), \\ D^{\alpha}e_{223} = f_2(z_1+a_3z_2+U_2) - g_2(c_1x_1 - x_2 - x_1x_3) \\ \qquad - h_3(y_1y_2 - b_2y_3(t-\tau)), \\ D^{\alpha}e_{331} = f_3(z_3(z_1-m_3)+b_3+U_3) - g_3(x_1x_2 - b_1x_3(t-\tau)) \\ \qquad - h_1(a_2(y_2-y_1)). \end{cases} \tag{44}$$

Here, we have the following similar results.

Theorem 3. *Multi-switching combination synchronization between the systems in Equations (24) and (27) and the system in Equation (30) can be achieved with the following controllers*

$$\begin{cases} U_1 = \dfrac{1}{f_1}\{(k_1-1)(f_1z_1 - g_1x_1 - h_2y_2) + g_1(a_1x_2 - (a_1+1)x_1 - x_1(t-\tau)) \\ \quad + h_2((c_2-a_2)y_1 - y_1y_3 + (c_2-1)y_2 - y_2(t-\tau)) - f_1(-(z_2+z_3) - 0.8z_1(t-\tau) - z_1)\}, \\ U_2 = \dfrac{1}{f_2}\{(k_2-1)(f_2z_2 - g_2x_2 - h_3y_3) + g_2(c_1x_1 - 2x_2 - x_1x_3 - x_2(t-\tau)) \\ \quad + h_3(y_1y_2 - (b_2+1)y_3(t-\tau) - y_3) - f_2(z_1 + (a_3-1)z_2 - z_2(t-\tau))\}, \\ U_3 = \dfrac{1}{f_3}\{(k_3-1)(f_3z_3 - g_3x_3 - h_1y_1) + g_3(x_1x_2 - (b_1+1)x_3(t-\tau) - x_3) \\ \quad + h_1(a_2(y_2-y_1) - y_1 - y_1(t-\tau)) - f_3(z_3(z_1-m_3) + b_3 - z_3 - z_3(t-\tau))\}. \end{cases} \tag{45}$$

Corollary 4. *(i) Supposing that $f_i \neq 0$, $g_i = 0$ and $h_i \neq 0$ $(i = 1,2,3)$, multi-switching hybrid projective synchronization between the systems in Equations (27) and (30) can be achieved with the following controllers*

$$\begin{cases} U_1 = \dfrac{1}{f_1}\{(k_1-1)(f_1z_1 - h_2y_2) + h_2((c_2-a_2)y_1 - y_1y_3 + (c_2-1)y_2 - y_2(t-\tau)) \\ \quad - f_1(-(z_2+z_3) - 0.8z_1(t-\tau) - z_1)\}, \\ U_2 = \dfrac{1}{f_2}\{(k_2-1)(f_2z_2 - h_3y_3) + h_3(y_1y_2 - (b_2+1)y_3(t-\tau) - y_3) \\ \quad - f_2(z_1 + (a_3-1)z_2 - z_2(t-\tau))\}, \\ U_3 = \dfrac{1}{f_3}\{(k_3-1)(f_3z_3 - h_1y_1) + h_1(a_2(y_2-y_1) - y_1 - y_1(t-\tau)) \\ \quad - f_3(z_3(z_1-m_3) + b_3 - z_3 - z_3(t-\tau))\}. \end{cases} \tag{46}$$

(ii) Supposing that $f_i \neq 0$, $g_i \neq 0$ and $h_i = 0$ $(i = 1,2,3)$, hybrid projective synchronization between the systems in Equations (24) and (30) can be achieved with the following controllers

$$\begin{cases} U_1 = \dfrac{1}{f_1}\{(k_1-1)(f_1z_1 - g_1x_1) + g_1(a_1x_2 - (a_1+1)x_1 - x_1(t-\tau)) \\ \quad - f_1(-(z_2+z_3) - 0.8z_1(t-\tau) - z_1)\}, \\ U_2 = \dfrac{1}{f_2}\{(k_2-1)(f_2z_2 - g_2x_2) + g_2(c_1x_1 - 2x_2 - x_1x_3 - x_2(t-\tau)) \\ \quad - f_2(z_1 + (a_3-1)z_2 - z_2(t-\tau))\}, \\ U_3 = \dfrac{1}{f_3}\{(k_3-1)(f_3z_3 - g_3x_3) + g_3(x_1x_2 - (b_1+1)x_3(t-\tau) - x_3) \\ \quad - f_3(z_3(z_1-m_3) + b_3 - z_3 - z_3(t-\tau))\}. \end{cases} \tag{47}$$

Corollary 5. *Supposing that $f_i \neq 0$, $g_i = 0$ and $h_i = 0$ $(i = 1, 2, 3)$, the system in Equation (30) can be stabilized to its equilibrium $O(0, 0, 0)$ with the following controllers*

$$\begin{cases} U_1 = \dfrac{1}{f_1}\{(k_1 - 1)(f_1 z_1) - f_1(-(z_2 + z_3) - 0.8z_1(t - \tau) - z_1)\}, \\ U_2 = \dfrac{1}{f_2}\{(k_2 - 1)(f_2 z_2) - f_2(z_1 + (a_3 - 1)z_2 - z_2(t - \tau))\}, \\ U_3 = \dfrac{1}{f_3}\{(k_3 - 1)(f_3 z_3) - f_3(z_3(z_1 - m_3) + b_3 - z_3 - z_3(t - \tau))\}. \end{cases} \tag{48}$$

The system parameters are given as $a_1 = 10$, $b_1 = \frac{8}{3}$, $c_1 = 28$, $a_2 = 35$, $b_2 = 3$, $c_2 = 28$, $a_3 = 0.4$, $b_3 = 0.2$, $m_3 = 10$, thus the systems in Equations (24), (27) and (30) exhibit chaotic behaviors, respectively. We assume $f_1 = f_2 = f_3 = 1$, $g_1 = g_2 = g_3 = 1$ and $h_1 = h_2 = h_3 = 1$, and the initial values are $(x_1(0), x_2(0), x_3(0)) = (-20, 2, 3)$, $(y_1(0), y_2(0), y_3(0)) = (7, 4.04, 20)$ and $(z_1(0), z_2(0), z_3(0)) = (1, 2, 40)$, respectively. Multi-switching combination synchronization between the systems in Equations (24), (27) and (30) can be realized with $K = diag\{-10, -10, -10\}$. Figure 4 illustrates synchronization errors e_{123}, e_{231}, e_{312}. Figure 5 shows synchronization states $x_2 + y_3$ vs. z_1, $x_3 + y_1$ vs. z_2 and $x_1 + y_2$ vs. z_3 of the drive systems in Equations (24) and (27) and the response system in Equation (30). Figure 6 displays synchronization errors e_{112}, e_{223}, and e_{331}. Figure 7 illustrates synchronization states $x_1 + y_2$ vs. z_1, $x_2 + y_3$ vs. z_2 and $x_3 + y_1$ vs. z_3 of the drive systems in Equations (24) and (27) and the response system in Equation (30). In Figures 4–7, we can see that the multi-switching combination synchronization errors converge to zero, i.e., the multi-switching combination synchronizations for Cases 1 and 2 are achieved, respectively.

The feedback gain matrix K is an important factor to affect the convergence of the error systems. With the increase of the absolute value of k_i, the convergence time will be shortened. Thus, we carried out one more simulation with $K = diag\{-40, -40, -40\}$. Figures 8 and 9 illustrate the synchronization errors for Cases 1 and 2, respectively. By comparing Figures 4 and 8, as well as Figures 6 and 9, it is easy to see that convergence times are shortened obviously.

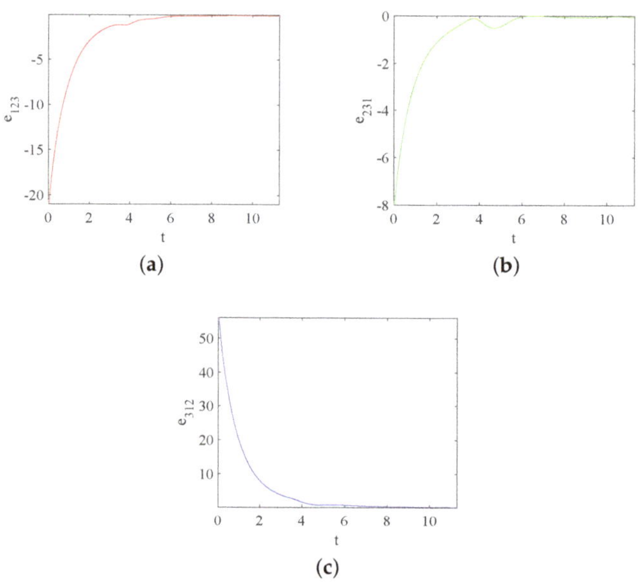

Figure 4. Synchronization errors among the systems in Equations (24), (27) and (30) with $k_1 = k_2 = k_3 = -10$: (**a**) e_{123}; (**b**) e_{231}; and (**c**) e_{312}.

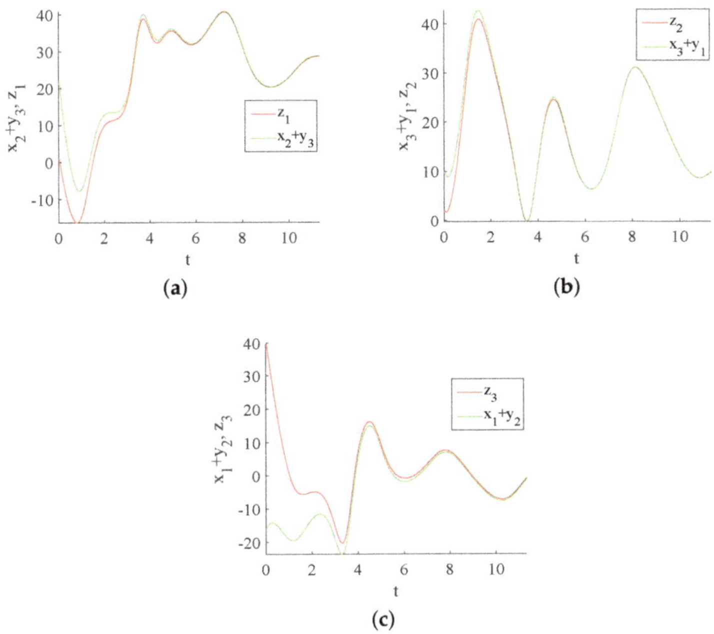

Figure 5. Responses for states between the systems in Equations (24) and (27) and the system in Equation (30): (**a**) $x_2 + y_3$ vs. z_1; (**b**) $x_3 + y_1$ vs. z_2; and (**c**) $x_1 + y_2$ vs. z_3.

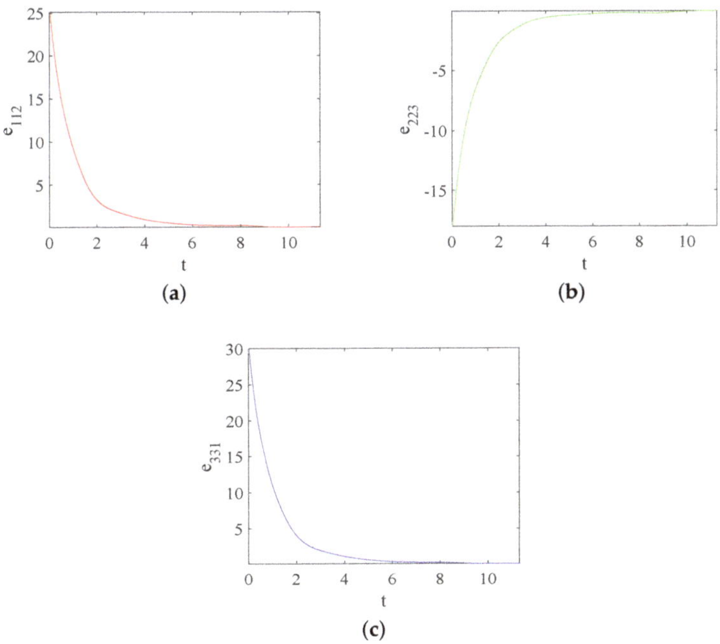

Figure 6. Synchronization errors among the systems in Equations (24), (27) and (30) with $k_1 = k_2 = k_3 = -10$: (**a**) e_{112}; (**b**) e_{223}; and (**c**) e_{331}.

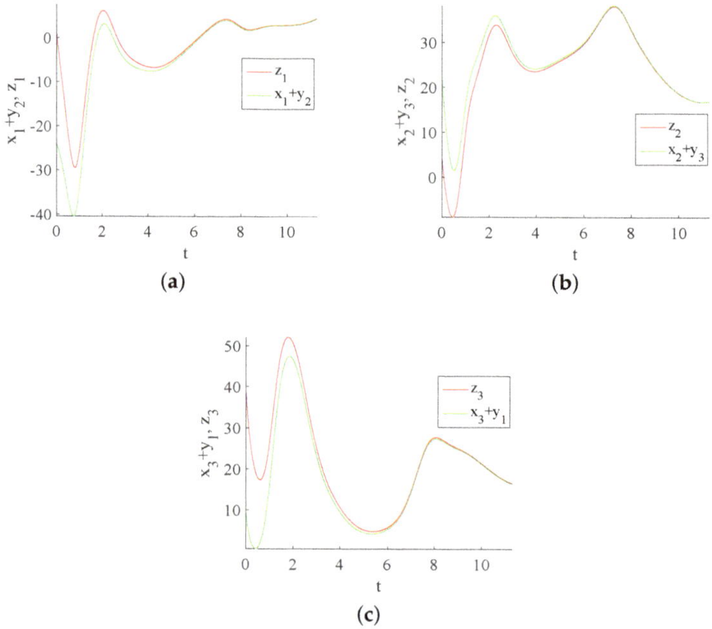

Figure 7. Responses for states between the systems in Equations (24) and (27) and the system in Equation (30): (**a**) $x_1 + y_2$ vs. z_1; (**b**) $x_2 + y_3$ vs. z_2; and (**c**) $x_3 + y_1$ vs. z_3.

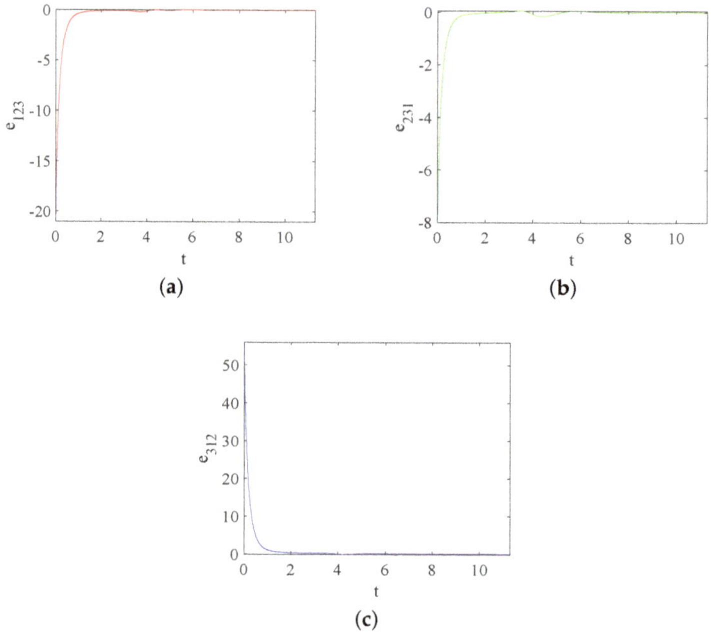

Figure 8. Synchronization errors among the systems in Equations (24), (27) and (30) with $k_1 = k_2 = k_3 = -40$: (**a**) e_{123}; (**b**) e_{231}; and (**c**) e_{312}.

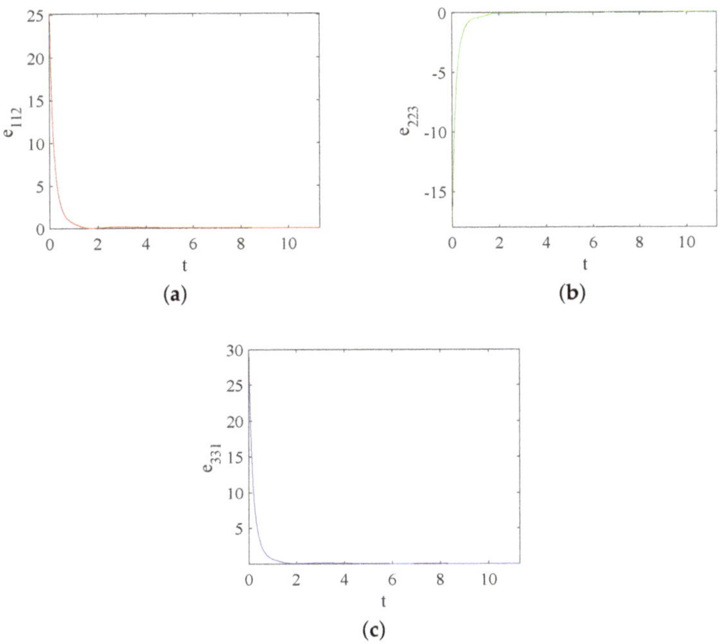

Figure 9. Synchronization errors among the systems in Equations (24), (27) and (30) with $k_1 = k_2 = k_3 = -40$: (**a**) e_{112}; (**b**) e_{223}; and (**c**) e_{331}.

5. Conclusions

We extended previous work [32] to investigate multi-switching combination synchronization among three non-identical fractional-order delayed systems by introducing time-delays. Based on the stability theory for linear fractional-order systems with multiple time-delays, we designed appropriate controllers to obtain multi-switching combination synchronization among three non-identical fractional-order delayed systems. The simulations are in accordance with the theoretical analysis.

On the one hand, when applying multi-switching combination synchronization of fractional-order delayed chaotic systems in secure communications, fractional-order and time-delay can enrich systems' dynamics. On the other hand, the origin information can be separated into two parts and embedded different parts in separate drive systems via combination synchronization scheme. Besides, because the switched states are unpredictable, this synchronization scheme can increase the security of the transmitted information in secure communication. Thus, the communication security will be enhanced, which makes multi-switching combination synchronization of fractional-order delayed chaotic systems able to find better applications in security communication.

Author Contributions: Methodology, X.Z.; Writing—Original Draft Preparation, B.L.; Writing—Review and Editing, Y.W.; and Funding Acquisition, X.Z.

Funding: This work was supported by the Natural Science Foundations of China under Grant No. 61463050, the NSF of Yunnan Province under Grant No. 2015FB113.

Acknowledgments: The authors sincerely thank the anonymous reviewers for their constructive comments and insightful suggestions.

Conflicts of Interest: The authors declare no conflict of interest.

References

1. Mandelbrot, B.B. *The Fractal Geometry of Nature*; WH Freeman: New York, NY, USA, 1983; Volume 173.
2. Boukal, Y.; Darouach, M.; Zasadzinski, M.; Radhy, N.E. Large-scale fractional-order systems: Stability analysis and their decentralised functional observers design. *IET Control. Theory Appl.* **2017**, *12*, 359–367. [CrossRef]
3. Caponetto, R. *Fractional Order Systems: Modeling and Control Applications*; World Scientific: Singapore, 2010; Volume 72.
4. Azar, A.T.; Vaidyanathan, S.; Ouannas, A. *Fractional Order Control and Synchronization of Chaotic Systems*; Springer: Berlin/Heidelberg, Germany, 2017; Volume 688.
5. Fahim, S.M.; Ahmed, S.; Imtiaz, S.A. Fractional order model identification using the sinusoidal input. *ISA Trans.* **2018**, *83*, 35–41. [CrossRef]
6. Lin, C.; Chen, B.; Wang, Q.G. Static output feedback stabilization for fractional-order systems in TS fuzzy models. *Neurocomputing* **2016**, *218*, 354–358. [CrossRef]
7. Bendoukha, S.; Ouannas, A.; Wang, X.; Khennaoui, A.A.; Pham, V.T.; Grassi, G.; Huynh, V. The Co-existence of Different Synchronization Types in Fractional-order Discrete-time Chaotic Systems with Non–identical Dimensions and Orders. *Entropy* **2018**, *20*, 710. [CrossRef]
8. Liu, Z.; Xia, T.; Wang, J. Fractional two-dimensional discrete chaotic map and its applications to the information security with elliptic-curve public key cryptography. *J. Vib. Control* **2018**, *24*, 4797–4824. [CrossRef]
9. Richard, J.P. Time-delay systems: An overview of some recent advances and open problems. *Automatica* **2003**, *39*, 1667–1694. [CrossRef]
10. Mircea, G.; Neamtu, M.; Opriş, D. *Uncertain, Stochastic and Fractional Dynamical Systems with Delay: Applications*; Lambert Academic Publishing: Saarbrücken, Germany, 2011.
11. Bhalekar, S.; Daftardar-Gejji, V. Fractional ordered Liu system with time-delay. *Commun. Nonlinear Sci. Numer. Simul.* **2010**, *15*, 2178–2191. [CrossRef]
12. Wang, Z.; Huang, X.; Shi, G. Analysis of nonlinear dynamics and chaos in a fractional order financial system with time delay. *Comput. Math. Appl.* **2011**, *62*, 1531–1539. [CrossRef]
13. Wang, S.; Yu, Y.; Wen, G. Hybrid projective synchronization of time-delayed fractional order chaotic systems. *Nonlinear Anal. Hybrid Syst.* **2014**, *11*, 129–138. [CrossRef]
14. Daftardar-Gejji, V.; Bhalekar, S.; Gade, P. Dynamics of fractional-ordered Chen system with delay. *Pramana* **2012**, *79*, 61–69. [CrossRef]
15. Song, X.; Song, S.; Li, B. Adaptive synchronization of two time-delayed fractional-order chaotic systems with different structure and different order. *Optik* **2016**, *127*, 11860–11870. [CrossRef]
16. Moaddy, K. Control and stability on chaotic convection in porous media with time delayed fractional orders. *Adv. Differ. Equ.* **2017**, *2017*, 311. [CrossRef]
17. Hu, J.B.; Zhao, L.D.; Xie, Z.G. Studying the intermittent stable theorem and the synchronization of a delayed fractional nonlinear system. *Chin. Phys. B* **2013**, *22*, 080506. [CrossRef]
18. Djennoune, S.; Bettayeb, M.; Al-Saggaf, U.M. Synchronization of fractional–order discrete–time chaotic systems by an exact delayed state reconstructor: Application to secure communication. *Int. J. Appl. Math. Comput. Sci.* **2019**, *29*, 179–194. [CrossRef]
19. Ding, D.; Qian, X.; Wang, N.; Liang, D. Synchronization and anti-synchronization of a fractional order delayed memristor-based chaotic system using active control. *Mod. Phys. Lett. B* **2018**, *32*, 1850142. [CrossRef]
20. Velmurugan, G.; Rakkiyappan, R. Hybrid projective synchronization of fractional-order chaotic complex nonlinear systems with time delays. *J. Comput. Nonlinear Dyn.* **2016**, *11*, 031016. [CrossRef]
21. He, S.; Sun, K.; Wang, H. Synchronisation of fractional-order time delayed chaotic systems with ring connection. *Eur. Phys. J. Spec. Top.* **2016**, *225*, 97–106. [CrossRef]
22. Runzi, L.; Yinglan, W.; Shucheng, D. Combination synchronization of three classic chaotic systems using active backstepping design. *Chaos Interdiscip. J. Nonlinear Sci.* **2011**, *21*, 043114. [CrossRef]
23. Sun, J.; Cui, G.; Wang, Y.; Shen, Y. Combination complex synchronization of three chaotic complex systems. *Nonlinear Dyn.* **2015**, *79*, 953–965. [CrossRef]
24. Xi, H.; Li, Y.; Huang, X. Adaptive function projective combination synchronization of three different fractional-order chaotic systems. *Optik* **2015**, *126*, 5346–5349. [CrossRef]

25. Zhou, X.; Jiang, M.; Huang, Y. Combination synchronization of three identical or different nonlinear complex hyperchaotic systems. *Entropy* **2013**, *15*, 3746–3761. [CrossRef]
26. Jiang, C.; Liu, S.; Wang, D. Generalized combination complex synchronization for fractional-order chaotic complex systems. *Entropy* **2015**, *17*, 5199–5217. [CrossRef]
27. Vincent, U.E.; Saseyi, A.; McClintock, P.V. Multi-switching combination synchronization of chaotic systems. *Nonlinear Dyn.* **2015**, *80*, 845–854. [CrossRef]
28. Zheng, S. Multi-switching combination synchronization of three different chaotic systems via nonlinear control. *Optik* **2016**, *127*, 10247–10258. [CrossRef]
29. Khan, A.; Khattar, D.; Prajapati, N. Adaptive multi switching combination synchronization of chaotic systems with unknown parameters. *Int. J. Dyn. Control* **2018**, *6*, 621–629. [CrossRef]
30. Ahmad, I.; Shafiq, M.; Al-Sawalha, M.M. Globally exponential multi switching-combination synchronization control of chaotic systems for secure communications. *Chin. J. Phys.* **2018**, *56*, 974–987. [CrossRef]
31. Hammami, S. Multi-switching combination synchronization of discrete-time hyperchaotic systems for encrypted audio communication. *IMA J. Math. Control. Inf.* **2018**, *36*, 583–602. [CrossRef]
32. Bhat, M.A.; Khan, A. Multi-switching combination synchronization of different fractional-order non-linear dynamical systems. *Int. J. Model. Simul.* **2018**, *38*, 254–261. [CrossRef]
33. Khan, A.; Bhat, M.A. Multi-switching combination–combination synchronization of non-identical fractional-order chaotic systems. *Math. Methods Appl. Sci.* **2017**, *40*, 5654–5667. [CrossRef]
34. Podlubny, I. *Fractional Differential Equations: An Introduction to Fractional Derivatives, Fractional Differential Equations, to Methods of Their Solution and Some of Their Applications*; Elsevier: Amsterdam, The Netherlands, 1998; Volume 198.
35. Kvitsinskii, A. Fractional integrals and derivatives: theory and applications. *Teor. Mater. Fiz.* **1993**, *3*, 397–414.
36. Deng, W.; Li, C.; Lü, J. Stability analysis of linear fractional differential system with multiple time delays. *Nonlinear Dyn.* **2007**, *48*, 409–416. [CrossRef]
37. Tang, J. Synchronization of different fractional order time-delay chaotic systems using active control. *Math. Probl. Eng.* **2014**, *2014*, 262151. [CrossRef]

applied sciences

MDPI

Article

Analytical Solutions of (2+Time Fractional Order) Dimensional Physical Models, Using Modified Decomposition Method

Hassan Khan [1], Umar Farooq [1], Rasool Shah [1], Dumitru Baleanu [2,3], Poom Kumam [4,5,*] and Muhammad Arif [1]

1 Department of Mathematics, Abdul Wali khan University, Mardan 23200, Pakistan; hassanmath@awkum.edu.pk (H.K.); umarfarooq0216@gmail.com (U.F.); rasoolshah@awkum.edu.pk (R.S.); marifmaths@awkum.edu.pk (M.A.)

2 Department of Mathematics, Faculty of Arts and Sciences, Cankaya University, 06530 Ankara, Turkey; dumitru@cankaya.edu.tr

3 Institute of Space Sciences, 077125 Magurele-Bucharest, Romania

4 Center of Excellence in Theoretical and Computational Science (TaCS-CoE) & Department of Mathematics, Faculty of Science, King Mongkut's University of Technology Thonburi (KMUTT), 126 Pracha Uthit Rd., Bang Mod, Thung Khru, Bangkok 10140, Thailand

5 Department of Medical Research, China Medical University Hospital, China Medical University, Taichung 40402, Taiwan

* Correspondence: poom.kum@kmutt.ac.th

Received: 24 October 2019; Accepted: 21 November 2019; Published: 23 December 2019

Abstract: In this article, a new analytical technique based on an innovative transformation is used to solve (2+time fractional-order) dimensional physical models. The proposed method is the hybrid methodology of Shehu transformation along with Adomian decomposition method. The series form solution is obtained by using the suggested method which provides the desired rate of convergence. Some numerical examples are solved by using the proposed method. The solutions of the targeted problems are represented by graphs which have confirmed closed contact between the exact and obtained solutions of the problems. Based on the novelty and straightforward implementation of the method, it is considered to be one of the best analytical techniques to solve linear and non-linear fractional partial differential equations.

Keywords: Shehu transformation; Adomian decomposition; analytical solution; Caputo derivatives; (2+time fractional-order) dimensional physical models

1. Introduction

Fractional calculus is considered to be a powerful tool for modeling complex phenomenon. Recently, the researchers have shown the greatest interest towards fractional calculus because of its numerous applications in different fields of sciences. Despite complicated background of fractional calculus, it came into being from simple question of L'Hospital. The first order represent slope of a function, what will it represent for fractional order $(\frac{1}{2})$? To find the answer of this question, the mathematicians have managed to open a new window of opportunities to improve the mathematical modeling of real world problems, which has given birth to many new questions and intriguing results. These newly established results have numerous implementation in many areas of engineering [1,2], such as fractional-order Buck master and diffusion problems [3], fractional-order telegraph model [4,5], fractional KdV-Burger-Kuramoto equation [6], fractal vehicular traffic flow [7], fractional Drinfeld-Sokolov-Wilson equation [8], fractional-order anomalous sub-diffusion model [9], fractional design of hepatitis B virus [10], fractional modeling chickenpox disease [11], fractional blood ethanol concentration model [12], fractional model for tuberculosis [13], fractional vibration equation [14], fractional Black-Scholes option pricing equations [15], fractionally damped beams [16], fractionally damped coupled system [17], fractional-order heat, wave and diffusion equations [18,19], fractional order pine wilt disease model [20], fractional diabetes model [21] etc.

Nowadays, the focus of the researchers is to develop different numerical and analytical techniques for the solution of fractional-order models. Therefore, different types of analytical and numerical methods have been developed and used for the solution of different fractional models. The analytical algorithm, the history of integral transform traced back to the time when Laplace started work an integral transform in 1780*s* and Joseph Fourier in 1822. Integral transformations are without question one of the most useful and effective methods in theoretical and applied mathematics, with numerous uses in quantum physics, mechanical engineering and several other areas of science. Moreover, the integral transform is used in chemistry, architecture, and other social sciences to evaluate various models [22]. In recent years, different integral transform such as Laplace transform [23–25] , Fourier transform [26,27], Hankel transform [28], Mellin transform [29], Z-transform [30], Wavelet transform [31], Elzaki transform [32,33], Mahgoub transform [34], Aboodh transform [35], Mohand transform [36], Sumudu transform [37,38], Hermite transform [39] etc have been used for the solution of different physical models.

Originality of the paper: In this article, we have applied a new analytical technique, which is based on generalization of sumudu and laplace transform with Adomian decomposition method (ADM) to solve (2+time fractional-order) dimensional physical models. In the present research we have analyzed the fractional view of some important physical problems by using Shehu decomposition method (SDM). Some important fractional-order problems are solved, which provide the best information about the targeted physical problems as compare to integer-order problems solution. The results of the integer-order problem are compared with the fractional-order problems. In conclusion, in the present research work, we provided and improved the existing physical models of integer-order by using the idea of fractional calculus. The modified mathematical models of fractional-order derivative are solved by using a new and sophisticated analytical method. Moreover, the proposed analytical method has provided the solutions of the problems that have a very close contact with the exact solutions of the problems. The methodology can be extended towards other fractional-order partial differential equations, that are frequently occurred in science and engineering.

The rest of the paper is organized as: In Section 2, we presented the basic definitions and theorem of the proposed method. In Section 3, we have discussed the implementation of proposed transformation. In Section 4 we evaluated the numerical examples by using the proposed technique and discussed the plots. In Section 5 we lastly summarized our results.

2. Preliminaries Concepts

In this section, we present some fundamental and appropriate definitions and preliminary concepts related to the fractional calculus and the Shehu transformation.

Definition 1. *Shehu transform*

Shehu transformation is new and similar to other integral transformation which is defined for functions of exponential order [40]. We take a function in the set A define by

$$A = \{u(\tau) : \exists, \rho_1, \rho_2 > 0, |u(\tau)| < Me^{\frac{|\tau|}{\rho_i}}, \ if \ \tau \in [0, \infty), \tag{1}$$

The Shehu transform which is represented by $S(.)$ for a function $u(\tau)$ is defined as

$$S\{u(\tau)\} = V(s, \mu) = \int_0^\infty u(\tau)e^{\frac{-s\tau}{\mu}} u(\tau)d\tau, \ \tau > 0, \ s > 0. \tag{2}$$

The Shehu transform of a function $u(\tau)$ is $V(s,\mu)$: then $u(\tau)$ is called the inverse of $V(s,\mu)$ which is expressed as

$$S^{-1}\{V(s,\mu)\} = u(\tau), \;\; for \;\; \tau \geq 0, \;\; S^{-1} is \;\; inverse \;\; Shehu \;\; transform. \tag{3}$$

Definition 2. *Shehu transform for nth derivatives*
The Shehu transformation for nth derivatives is defined as [40]

$$S\left\{u^{(n)}(\tau)\right\} = \frac{s^n}{\mu^n}V(s,\mu) - \sum_{k=0}^{n-1}\left(\frac{s}{\mu}\right)^{n-k-1} u^{(k)}(0). \tag{4}$$

Definition 3. *Caputo operator of fractional partial derivative*
The fractional Caputo operator is represented as [41]

$$D_\tau^\beta f(\tau) = \begin{cases} \frac{\partial^n f(\tau)}{\partial \tau^n}, & \beta = n \in N, \\ \frac{1}{\Gamma(n-\beta)}\int_0^\tau (\tau-\phi)^{n-\beta-1} f^{(n)}(\phi)\partial\phi, & n-1 < \beta \leq n, \quad n \in N. \end{cases} \tag{5}$$

Definition 4. *Shehu transform for fractional order derivatives*
The Shehu transformation for the fractional order derivatives is expressed as

$$S\left\{u^{(\beta)}(\tau)\right\} = \frac{s^\beta}{\mu^\beta}V(s,\mu) - \sum_{k=0}^{n-1}\left(\frac{s}{\mu}\right)^{\beta-k-1} u^{(k)}(0), \quad 0 < \beta \leq n, \tag{6}$$

In Table 1 show different special functions of Shehu transformation.

Table 1. The Shehu transform of some special functions.

Functional Form	Shehu Transform Form
1	$\frac{u}{s}$
t	$\frac{u^2}{s^2}$
e^τ	$\frac{u}{s-au}$
$\sin(\tau)$	$\frac{u^2}{s^2+u^2}$
$\cos(\tau)$	$\frac{us}{s^2+u^2}$
$\frac{\tau^n}{n!}$ for $n = 0,1,2,\cdots$	$(\frac{u}{s})^{n+1}$
$\frac{\tau^n}{\Gamma(n+1)}$ for $n = 0,1,2,\cdots$	$(\frac{u}{s})^{n+1}$

Theorem 1. *If the function $u(\tau)$ is piecewise continues at every finite interval of $0 \leq \tau \leq \beta$ and of exponential order α for $\tau > \beta$, then there's the Shehu transform $u(s,\mu)$ [40].*

Proof. For any natural number β, we deduct algebraically:

$$\int_0^\infty exp(-\frac{s\tau}{\mu})u(\tau)d\tau = \int_0^\beta exp(-\frac{s\tau}{\mu})u(\tau)d\tau + \int_\beta^\infty exp(-\frac{s\tau}{\mu})u(\tau)d\tau, \tag{7}$$

since the function $u(\tau)$ continues in a piecewise manner at every finite interval $0 \leq \tau \leq \beta$, there's the first integral on the right hand side. We suggest the following situation to validate this statement,

$$
\begin{aligned}
|\int_{\alpha}^{\infty} exp(-\frac{s\tau}{\mu})u(\tau)d\tau| &\leq \int_{\alpha}^{\infty} |exp(-\frac{s\tau}{\mu})u(\tau)|d\tau \\
&\leq \int_{\alpha}^{\infty} exp(-\frac{s\tau}{\mu})|u(\tau)|d\tau \\
&\leq \int_{\alpha}^{\infty} exp(-\frac{s\tau}{\mu})Nexp(\beta\tau)d\tau \\
&= N\int_{\alpha}^{\infty} exp(-\frac{(s-\beta u)\tau}{u})d\tau \\
&= -\frac{Nu}{(s-\beta u)}\lim_{\gamma\longrightarrow\infty}\left[exp(-\frac{(s-\beta u)\tau}{u})d\tau\right]_0^{\gamma} \\
&= \frac{Nu}{(s-\beta u)}.
\end{aligned} \tag{8}
$$

The proof is complete. □

3. Implementation of Shehu Transform

In this section, we have considered a time fractional (2+time fractional-order) dimensional physical model in the form

$$u_{\tau}^{\beta}(\Im,\Re,\tau) = \kappa u_{\Im\Im}(\Im,\Re,\tau) + \pounds u(\Im,\Re,\tau) + \aleph u(\Im,\Re,\tau), \quad \beta \in [1,2] \tag{9}$$

with initial condition

$$u(\Im,\Re,0) = u(\Im,\Re), \tag{10}$$

while κ is a non-linear operator and £ linear operator.

Applying the Shehu transform to both sides of the Equation (9) we obtain

$$S\left\{u_{\tau}^{\beta}(\Im,\Re,\tau)\right\} = S\left\{\kappa u_{\Im\Im}(\Im,\Re,\tau) + \pounds u(\Im,\Re,\tau) + \aleph u(\Im,\Re,\tau)\right\}, \quad \beta \in [1,2]. \tag{11}$$

Using the differential property of Shehu transformation we have,

$$\frac{s^{\beta}}{\mu^{\beta}}\left\{V(s,\mu) - \frac{\mu}{s}u(0) - \frac{\mu^2}{s^2}u^{'}(0)\right\} = S\left\{\kappa u_{\Im\Im}(\Im,\Re,\tau) + \pounds u(\Im,\Re,\tau) + \aleph u(\Im,\Re,\tau)\right\}. \tag{12}$$

Simplifying Equation (12), we obtain

$$V(s,\mu) = +\frac{\mu^{\beta}}{s^{\beta}}S\left\{\kappa u_{\Im\Im}(\Im,\Re,\tau) + \pounds u(\Im,\Re,\tau) + \aleph u(\Im,\Re,\tau)\right\} + \frac{\mu}{s}u(0) + \frac{\mu^2}{s^2}u^{'}(0). \tag{13}$$

Applying the inverse Shehu transformation, we get

$$u(\Im,\Re,\tau) = S^{-1}\left\{\frac{\mu^{\beta}}{s^{\beta}}S\left\{\kappa u_{\Im\Im}(\Im,\Re,\tau) + \pounds u(\Im,\Re,\tau) + \aleph u(\Im,\Re,\tau)\right\}\right\} + u(0) + \tau u^{'}(0). \tag{14}$$

The nonlinear term $\aleph u(\Im,\Re,\tau)$ is evaluated by using the procedure of Adomian polynomial decomposition given by

$$\aleph u(\Im,\Re,\tau) = \sum_{m=0}^{\infty} A_m(u_0,u_1,\cdots), \; m = 0,1,\cdots \tag{15}$$

where,

$$A_m(u_0,u_1,\cdots) = \frac{1}{m!}\left[\frac{d^m}{d\lambda^m}\aleph\left(\sum_{i=0}^{\infty}\lambda^i u_i\right)\right]_{\lambda=0}, \quad m > 0. \tag{16}$$

With the help of Equation (16), Equation (15) can be written as

$$u(\Im,\Re,\tau)=S^{-1}\left\{\frac{\mu^{\beta}}{s^{\beta}}S\left\{\kappa u_{\Im\Im}(\Im,\Re,\tau)+\pounds u(\Im,\Re,\tau)+\sum_{m=0}^{\infty}A_m\right\}\right\}+u(0)+\tau u^{'}(0). \tag{17}$$

Finally, we obtain the recursive relation as

$$\begin{aligned}&u_0(\Im,\Re,\tau)=u(0)+\tau u^{'}(0),\ \ m=0\\&u_m(\Im,\Re,\tau)=S^{-1}\left\{\frac{\mu^{\beta}}{s^{\beta}}S\left\{\kappa u_{(m-1)\Im\Im}(\Im,\Re,\tau)+\pounds u_{(m-1)}(\Im,\Re,\tau)+\aleph u_{(m-1)}(\Im,\Re,\tau)+\sum_{m=0}^{\infty}A_m\right\}\right\},\ \ m\geq 1.\end{aligned} \tag{18}$$

4. Applications and Discussion

Example 1. *Consider the (2+time fractional-order) dimensional hyperbolic wave model:*

$$u_{\tau}^{\beta}(\Im,\Re,\tau)=\frac{1}{12}\Im^2u_{\Im\Im}(\Im,\Re,\tau)+\frac{1}{12}\Re^2u_{\Re\Re}(\Im,\Re,\tau),\quad \beta\in(1,2) \tag{19}$$

with initial conditions

$$u(\Im,\Re,0)=\Im^4,\qquad u_{\tau}(\Im,\Re,0)=\Re^4. \tag{20}$$

If $\beta=2$, then the exact solution of Equation (19) is

$$u(\Im,\Re,\tau)=\Im^4\sinh(t)+\Re^4\cosh(t), \tag{21}$$

Taking the Shehu transform of Equation (19) we obtain

$$\frac{s^{\beta}}{\mu^{\beta}}\left\{V(s,\mu)-\frac{\mu}{s}u(0)-\frac{\mu^2}{s^2}u^{'}(0)\right\}=S\left\{\frac{1}{12}\Im^2u_{\Im\Im}(\Im,\Re,\tau)+\frac{1}{12}\Re^2u_{\Re\Re}(\Im,\Re,\tau)\right\}. \tag{22}$$

Simplifying Equation (22), we get

$$V(s,\mu)=\frac{\mu^{\beta}}{s^{\beta}}S\left\{\frac{1}{12}\Im^2u_{\Im\Im}(\Im,\Re,\tau)+\frac{1}{12}\Re^2u_{\Re\Re}(\Im,\Re,\tau)\right\}+\frac{\mu}{s}u(0)+\frac{\mu^2}{s^2}u^{'}(0). \tag{23}$$

Applying inverse Shehu transform, we get

$$u(\Im,\Re,\tau)=u(0)+u^{'}(0)\tau+S^{-1}\left\{\frac{\mu^{\beta}}{s^{\beta}}S\left\{\frac{1}{12}\Im^2u_{\Im\Im}(\Im,\Re,\tau)+\frac{1}{12}\Re^2u_{\Re\Re}(\Im,\Re,\tau)\right\}\right\}. \tag{24}$$

Thus we get the following recursive scheme

$$u_0(\Im,\Re,\tau)=u(0)+u^{'}(0)\tau=\Im^4+\Re^4\tau, \tag{25}$$

$$u_{m+1}(\Im,\Re,\tau)=S^{-1}\left\{\frac{\mu^{\beta}}{s^{\beta}}S\left\{\frac{1}{12}\Im^2u_{m\Im\Im}(\Im,\Re,\tau)+\frac{1}{12}\Re^2u_{m\Re\Re}(\Im,\Re,\tau)\right\}\right\}. \tag{26}$$

Using Equation (26), for $m=0,1,2,3,\cdots$ we get the following values

$$\begin{aligned}
u_1(\Im,\Re,\tau) &= \Im^4\frac{\tau^\beta}{\beta!}+\Re^4\frac{\tau^{\beta+1}}{(\beta+1)!}',\\
u_2(\Im,\Re,\tau) &= \Im^4\frac{\tau^{2\beta}}{(2\beta)!}+\Re^4\frac{\tau^{2\beta+1}}{(2\beta+1)!}',\\
u_3(\Im,\Re,\tau) &= \Im^4\frac{\tau^{3\beta}}{(3\beta)!}+\Re^4\frac{\tau^{3\beta+1}}{(3\beta+1)!}',\\
u_4(\Im,\Re,\tau) &= \Im^4\frac{\tau^{4\beta}}{(4\beta)!}+\Re^4\frac{\tau^{4\beta+1}}{(4\beta+1)!}',\\
&\vdots
\end{aligned} \tag{27}$$

Now using the values of $u_0, u_1, u_2, u_3, \cdots$, we get Shehu transformation solution for example 1

$$u(\Im,\Re,\tau) = \Im^4+\Re^4\tau+\Im^4\frac{\tau^\beta}{\beta!}+\Re^4\frac{\tau^{\beta+1}}{(\beta+1)!}+\Im^4\frac{\tau^{2\beta}}{(2\beta)!}+\Re^4\frac{\tau^{2\beta+1}}{(2\beta+1)!}+\Im^4\frac{\tau^{3\beta}}{(3\beta)!}+ \\ \Re^4\frac{t^{3\beta+1}}{(3\beta+1)!}+\Im^4\frac{\tau^{4\beta}}{(4\beta)!}+\Re^4\frac{\tau^{4\beta+1}}{(4\beta+1)!}+\cdots \tag{28}$$

After simplification, we get

$$u(\Im,\Re,\tau) = \Im^4\left\{1+\frac{\tau^\beta}{\beta!}+\frac{\tau^{2\beta}}{(2\beta)!}+\frac{\tau^{3\beta}}{(3\beta)!}+\frac{\tau^{4\beta}}{(4\beta)!}+\cdots\right\}+\Re^4\left\{\tau+\frac{\tau^{\beta+1}}{(\beta+1)!}+\frac{\tau^{2\beta+1}}{(2\beta+1)!}+ \right. \\ \left. \frac{\tau^{3\beta+1}}{(3\beta+1)!}+\frac{\tau^{4\beta+1}}{(4\beta+1)!}+\cdots\right\}. \tag{29}$$

In particular, when $\beta\to 2$, the analytical solution of Shehu transform become as

$$u(\Im,\Re,\tau) = \Im^4\left\{1+\frac{\tau^2}{2!}+\frac{\tau^4}{(4)!}+\frac{\tau^6}{(6)!}+\frac{\tau^8}{(8)!}+\cdots\right\}+\Re^4\left\{\tau+\frac{\tau^3}{(3)!}+\frac{\tau^5}{(5)!}+\frac{\tau^7}{(7)!}+\frac{\tau^9}{(9)!}+\cdots\right\}, \tag{30}$$

which provide the close form solution as

$$u(\Im,\Re,\tau) = \Im^4\cosh(\tau)+\Re^4\sinh(\tau). \tag{31}$$

Figures 1 and 2 represent the exact and analytical solutions of Example 1. The solutions-graphs have confirmed the closed contact between the exact solution and the analytical solution obtained by the proposed method. In Figure 3, the solution of Example 1 are calculated at different fractional-order β of the derivative. It is investigated that the solutions at different fractional-orders β are convergent to an integer-order solution of Example 1. Figure 4 represent the solution verses time graph for Example 1. It is observed that as the time fractional-order varies toward time integer-order, the time fractional-order solutions also approaches to the solution of an integer-order problem of Example 1. All the above solution analysis of Example 1 indicate that SDM is an efficient and effective method to solve fractional-order partial differential equations that are frequently arising in science and engineering.

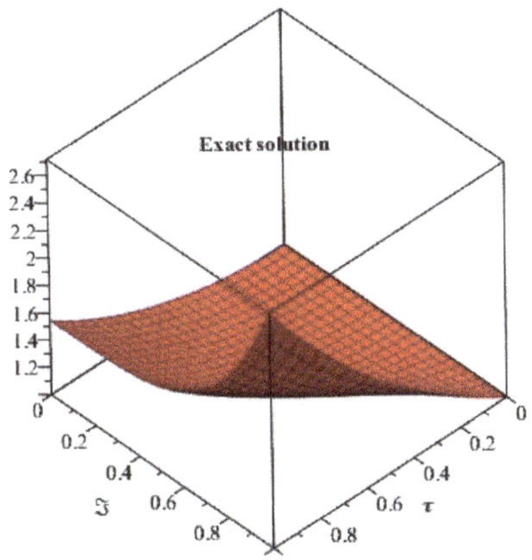

Figure 1. Represents the exact solution of Example 1 at $\beta = 2$.

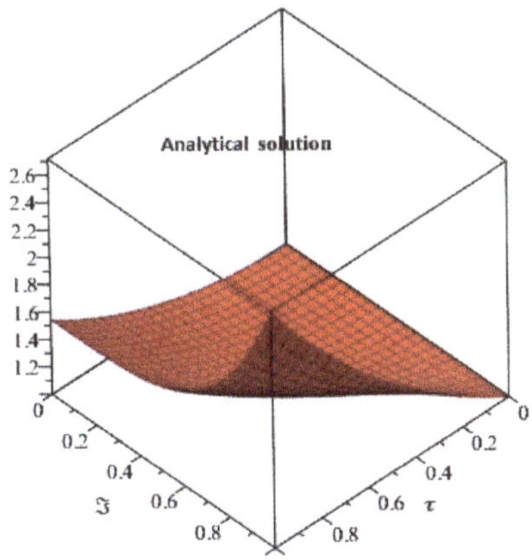

Figure 2. Represents the analytical solution of Example 1 at $\beta = 2$.

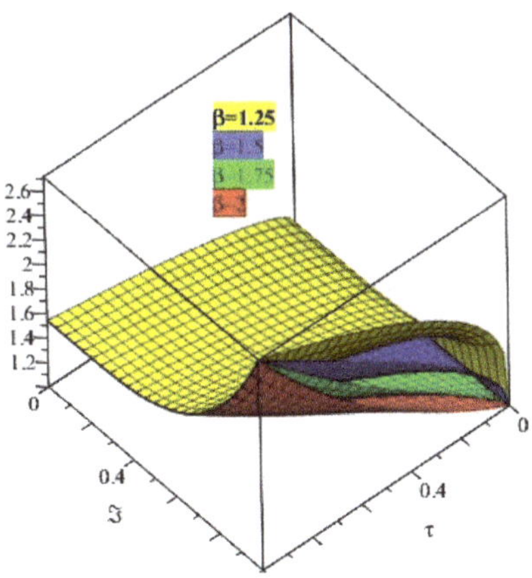

Figure 3. Represents the solution at different fractional order of Example 1.

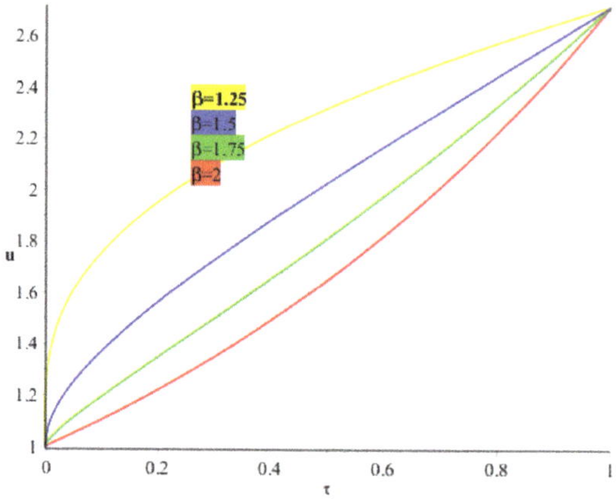

Figure 4. Represents the solution at different fractional order of Example 1.

In Table 2, the solutions of Shehu transform decomposition method (SDM) and Adomian decomposition method (ADM) are compared with each other. The comparison has shown that the solutions of proposed method are in strong agreement with the solution of ADM.

Table 2. Comparison of SDM and ADM [42] of Example 1 at $\tau = 0.1$.

		SDM (m = 5)	SDM (m = 3)	SDM (m= 5)	ADM (m = 5)	AE of SDM
$\Im$	$\Re$	$\beta = 1.75$	$\beta = 2$	$\beta = 2$	$\beta = 2$	$\beta = 2$
1	1	1.111568974	1.105195833	1.10519608	1.10519609	2.51×10^{-5}
2	2	17.78510358	17.68313333	17.6831373	17.6831374	4.02×10^{-4}
3	3	90.03708688	89.52086250	89.5208829	89.5208828	2.03×10^{-3}
4	4	284.5616573	282.9301334	282.930198	282.930199	6.44×10^{-3}
5	5	694.7306086	690.7473959	690.747553	690.747552	1.57×10^{-2}

Example 2. *Consider the (2+time fractional-order) dimensional Heat model:*

$$u_{\tau}^{\beta}(\Im, \Re, \tau) = u_{\Im\Im}(\Im, \Re, \tau) + u_{\Re\Re}(\Im, \Re, \tau), \quad \beta \in (0, 1] \tag{32}$$

with initial condition

$$u(\Im, \Re, 0) = \sin(\Im)\cos(\Re). \tag{33}$$

If $\beta = 1$, then the exact solution of Equation (32) is

$$u(\Im, \Re, \tau) = e^{-2\tau}\sin(\Im)\cos(\Re). \tag{34}$$

Taking Shehu transform of Equation (32)

$$\frac{s^{\beta}}{\mu^{\beta}}\left\{V(s,\mu) - \frac{\mu}{s}u(0)\right\} = S\left\{u_{\Im\Im}(\Im, \Re, \tau) + u_{\Re\Re}(\Im, \Re, \tau)\right\}, \tag{35}$$

Simplifying Equation (35), we get as

$$V(s,\mu) = \frac{\mu}{s}u(0) + \frac{\mu^{\beta}}{s^{\beta}}S\left\{u_{\Im\Im}(\Im, \Re, \tau) + u_{\Re\Re}(\Im, \Re, \tau)\right\}. \tag{36}$$

Applying inverse Shehu transform, we get

$$u(\Im,\Re,\tau)=u(0)+S^{-1}\left\{\frac{\mu^{\beta}}{s^{\beta}}S\left\{u_{\Im\Im}(\Im,\Re,\tau)+u_{\Re\Re}(\Im,\Re,\tau)\right\}\right\}. \tag{37}$$

Thus we get the following recursive scheme

$$u_0(\Im,\Re,\tau)=u(0)=\sin(\Im)\cos(\Re), \tag{38}$$

$$u_{m+1}(\Im,\Re,\tau)=S^{-1}\left\{\frac{\mu^{\beta}}{s^{\beta}}S\left\{u_{m\Im\Im}(\Im,\Re,\tau)+u_{m\Re\Re}(\Im,\Re,\tau)\right\}\right\}, \tag{39}$$

Using Equation (39), for $m=0,1,2,3,\cdots$ *we get the following values*

$$\begin{aligned}
u_1(\Im,\Re,\tau)&=-2\sin(\Im)\cos(\Re)\frac{\tau^{\beta}}{(\beta)!},\\
u_2(\Im,\Re,\tau)&=4\sin(\Im)\cos(\Re)\frac{\tau^{2\beta}}{(2\beta)!},\\
u_3(\Im,\Re,\tau)&=-8\sin(\Im)\cos(\Re)\frac{\tau^{3\beta}}{(3\beta)!},\\
u_4(\Im,\Re,\tau)&=16\sin(\Im)\cos(\Re)\frac{\tau^{4\beta}}{(4\beta)!},\\
&\vdots
\end{aligned} \tag{40}$$

Now using the values of $u_0,u_1,u_2,u_3,\cdots$, *we get Shehu transformation solution for example 2*

$$\begin{aligned}
u(\Im,\Re,\tau)=&\sin(\Im)\cos(\Re)-2\sin(\Im)\cos(\Re)\frac{\tau^{\beta}}{(\beta)!}+4\sin(\Im)\cos(\Re)\frac{\tau^{2\beta}}{(2\beta)!}+\\
&-8\sin(\Im)\cos(\Re)\frac{\tau^{3\beta}}{(3\beta)!}+16\sin(\Im)\cos(\Re)\frac{\tau^{4\beta}}{(4\beta)!}+\cdots
\end{aligned} \tag{41}$$

After simplification, we get

$$u(\Im,\Re,\tau)=\sin(\Im)\cos(\Re)\left\{1-2\frac{\tau^{\beta}}{(\beta)!}+4\frac{\tau^{2\beta}}{(2\beta)!}+-8\frac{\tau^{3\beta}}{(3\beta)!}+16\frac{\tau^{4\beta}}{(4\beta)!}+\cdots\right\}, \tag{42}$$

which converge to the solution

$$u(\Im,\Re,\tau)=\sin(\Im)\cos(\Re)E_{\beta}(-2\tau^{\beta}), \tag{43}$$

For particular case $\beta\to1$, *the Shehu transform solution become as*

$$u(\Im,\Re,\tau)=\sin(\Im)\cos(\Re)e^{-2\tau}. \tag{44}$$

Figures 5 and 6 show the exact and analytical solution of Example 2 respectively. The graphical representation have confirmed the closed contact of the obtained solution with the exact solution of Example 2. Similarly, Figures 7 and 8 represents the fractional-order solution of Example 2 for two and three space. Both graphs support the convergence phenomena of fractional-order problems to an integer-order problem of Example 2.

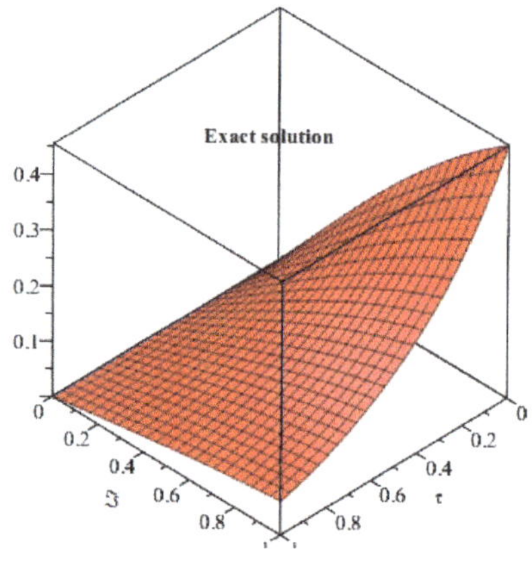

Figure 5. Represents the exact solution of Example 2 at $\beta = 1$.

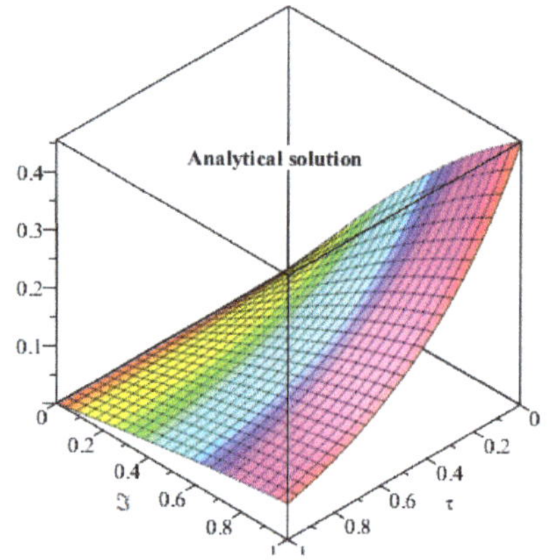

Figure 6. Represents the analytical solution of Example 2 at $\beta = 1$.

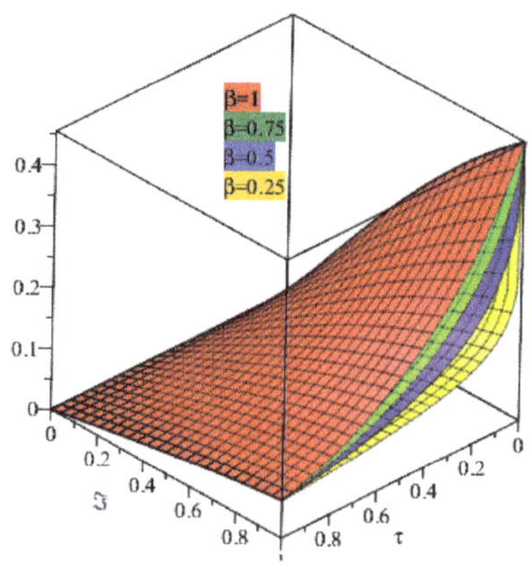

Figure 7. $u(\Im, \Re, \tau)$ Represents the solution at different fractional order of Example 2.

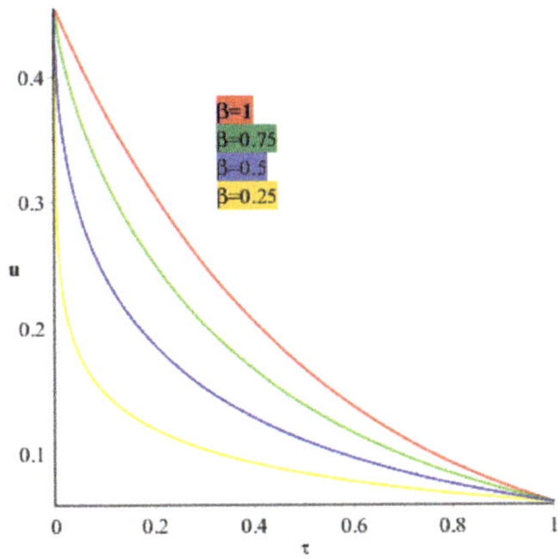

Figure 8. $u(\Im, \Re, \tau)$ Represents the solution at different fractional order of Example 2.

Example 3. *Consider the* $(2 + time fractional)$ *dimensional diffusion model:*

$$u_{\tau}^{\beta}(\Im, \Re, \tau) = u_{\Im\Im}(\Im, \Re, \tau) + u_{\Re\Re}(\Im, \Re, \tau), \quad \beta \in (0, 1] \tag{45}$$

with the initial condition

$$u(\Im, \Re, 0) = e^{\Im+\Re}. \tag{46}$$

If $\beta = 1$, *then the exact solution of Equation (45) is*

$$u(\Im, \Re, \tau) = e^{\Im+\Re+2\tau} \tag{47}$$

Taking Shehu transform of Equation (45)

$$\frac{s^{\beta}}{\mu^{\beta}} \left\{ V(s, \mu) - \frac{\mu}{s} u(0) \right\} = S \left\{ u_{\Im\Im}(\Im, \Re, \tau) + u_{\Re\Re}(\Im, \Re, \tau) \right\}. \tag{48}$$

Simplifying Equation (46), we get as

$$V(s, \mu) = \frac{\mu}{s} u(0) + \frac{\mu^{\beta}}{s^{\beta}} S \left\{ u_{\Im\Im}(\Im, \Re, \tau) + u_{\Re\Re}(\Im, \Re, \tau) \right\}. \tag{49}$$

Applying inverse operator of Shehu transform, we get

$$u(\Im, \Re, \tau) = u(0) + S^{-1} \left\{ \frac{\mu^{\beta}}{s^{\beta}} S \left\{ u_{\Im\Im}(\Im, \Re, \tau) + u_{\Re\Re}(\Im, \Re, \tau) \right\} \right\}. \tag{50}$$

Thus we get the following recursive scheme

$$u_0(\Im, \Re, \tau) = u(0) = e^{\Im+\Re},$$

$$u_{m+1}(\Im, \Re, \tau) = S^{-1} \left\{ \frac{\mu^{\beta}}{s^{\beta}} S \left\{ u_{m\Im\Im}(\Im, \Re, \tau) + u_{m\Re\Re}(\Im, \Re, \tau) \right\} \right\}, \tag{51}$$

Using Equation (51), for $m = 0, 1, 2, 3, \cdots$ *we get the following values*

$$\begin{aligned} u_1(\Im,\Re,\tau) &= 2e^{\Im+\Re}\frac{\tau^{\beta}}{(\beta)!}, \\ u_2(\Im,\Re,\tau) &= 4e^{\Im+\Re}\frac{\tau^{2\beta}}{(2\beta)!}, \\ u_3(\Im,\Re,\tau) &= 8e^{\Im+\Re}\frac{\tau^{3\beta}}{(3\beta)!}, \\ u_4(\Im,\Re,\tau) &= 16e^{\Im+\Re}\frac{\tau^{4\beta}}{(4\beta)!}, \\ &\vdots \end{aligned} \tag{52}$$

Now using the values of $u_0, u_1, u_2, u_3, \cdots$, we get Shehu transformation solution for Example 3

$$u(\Im,\Re,\tau) = e^{\Im+\Re} + 2e^{\Im+\Re}\frac{\tau^{\beta}}{(\beta)!} + 4e^{\Im+\Re}\frac{\tau^{2\beta}}{(2\beta)!} + 8e^{\Im+\Re}\frac{\tau^{3\beta}}{(3\beta)!} + 16e^{\Im+\Re}\frac{\tau^{4\beta}}{(4\beta)!} + \cdots. \tag{53}$$

After simplification, we get

$$u(\Im,\Re,\tau) = \sin(\Im)\cos(\Re)\left\{1 + 2\frac{\tau^{\beta}}{(\beta)!} + 4\frac{\tau^{2\beta}}{(2\beta)!} + 8\frac{\tau^{3\beta}}{(3\beta)!} + 16\frac{\tau^{4\beta}}{(4\beta)!} + \cdots\right\}. \tag{54}$$

The close form solution become as

$$u(\Im,\Re,\tau) = \sin(\Im)\cos(\Re)E_{\beta}(2\tau^{\beta}). \tag{55}$$

When $\beta \to 1$ the calculated result provide the exact solution in the close form

$$u(\Im,\Re,\tau) = \sin(\Im)\cos(\Re)e^{2\tau}. \tag{56}$$

Figures 9 and 10 show the exact and analytical solutions of Example 3. Both figures are almost coincident confirming the close contact of both exact and obtained solution. Figures 11 the SDM solutions at different fractional-order β are calculated for Example 3. The convergence phenomena of fractional-order solution towards exact solution is observed. The method is found to be very simple and straightforward to solve fractional-order different equations.

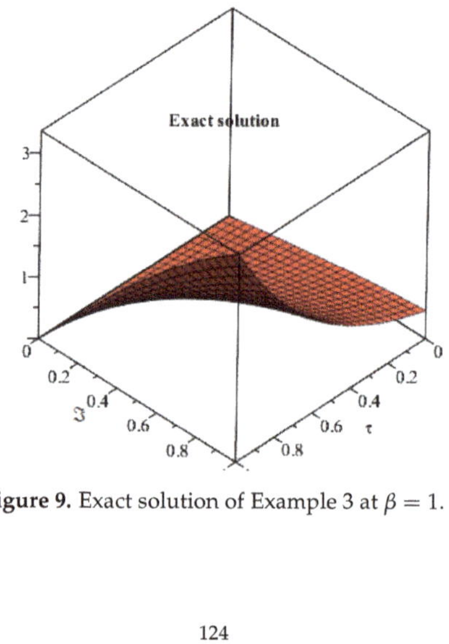

Figure 9. Exact solution of Example 3 at $\beta = 1$.

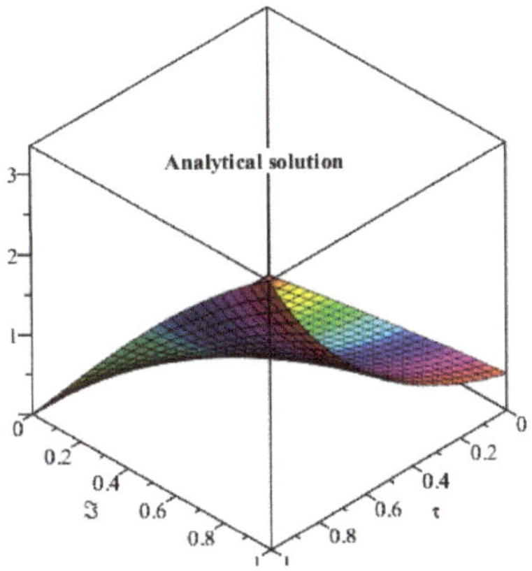

Figure 10. Represents the analytical solution of Example 3 at $\beta = 1$.

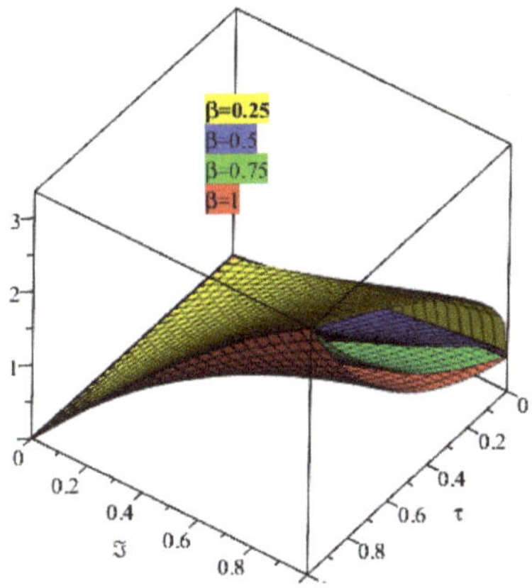

Figure 11. The solution graph at different fractional order β.

Example 4. *Consider the* $(2 + time fractional)$ *dimensional telegraph model:*

$$u_\tau^\beta(\Im, \Re, \tau) = \frac{1}{2}u_{\Im\Im}(\Im, \Re, \tau) + \frac{1}{2}u_{\Re\Re}(\Im, \Re, \tau) - 2u_t(\Im, \Re, \tau) - u(\Im, \Re, \tau), \quad \beta \in (1, 2], \tag{57}$$

with initial conditions

$$u(\Im, \Re, 0) = \sinh(\Im)\sinh(\Re), \qquad u_\tau(\Im, \Re, 0) = -2\sinh(\Im)\sinh(\Re). \tag{58}$$

If $\beta = 2$, *then the exact solution of Equation (57) is*

$$u(\Im, \Re, \tau) = \sinh(\Im)\sinh(\Re)e^{-2\tau}. \tag{59}$$

Taking Shehu transform of Equation (57)

$$\frac{s^\beta}{\mu^\beta}\left\{V(s,\mu) - \frac{\mu}{s}u(0) - \frac{\mu^2}{s^2}u'(0)\right\} = S\left\{\frac{1}{2}u_{\Im\Im}(\Im, \Re, \tau) + \frac{1}{2}u_{\Re\Re}(\Im, \Re, \tau) - 2u_t(\Im, \Re, \tau) - u(\Im, \Re, \tau)\right\}, \tag{60}$$

Simplifying Equation (60), we get as

$$V(s,\mu) = \frac{\mu^\beta}{s^\beta}S\left\{\frac{1}{2}u_{\Im\Im}(\Im, \Re, \tau) + \frac{1}{2}u_{\Re\Re}(\Im, \Re, \tau) - 2u_t(\Im, \Re, \tau) - u(\Im, \Re, \tau)\right\} + \frac{\mu}{s}u(0) + \frac{u^2}{s^2}u'(0), \tag{61}$$

Applying inverse of Shehu transform, we get

$$u(\Im,\Re,\tau)=u(0)+\tau u'(0)+S^{-1}\left\{\frac{\mu^{\beta}}{s^{\beta}}S\left\{\frac{1}{2}u_{\Im\Im}(\Im,\Re,\tau)+\frac{1}{2}u_{\Re\Re}(\Im,\Re,\tau)-2u_{t}(\Im,\Re,\tau)-u(\Im,\Re,\tau)\right\}\right\}. \tag{62}$$

Thus we get the following recursive scheme

$$\begin{aligned} u_0(\Im,\Re,\tau) &= u(0)+\tau u'(0) = \\ &\sinh(\Im)\sinh(\Re)-2t\sinh(\Im)\sinh(\Re), \end{aligned}$$

$$u_{m+1}(\Im,\Re,\tau)=S^{-1}\left\{\frac{\mu^{\beta}}{s^{\beta}}S\left\{\frac{1}{2}u_{m\Im\Im}(\Im,\Re,\tau)+\frac{1}{2}u_{m\Re\Re}(\Im,\Re,\tau)-2u_{m\tau}(\Im,\Re,\tau)-u_{m}(\Im,\Re,\tau)\right\}\right\}, \tag{63}$$

Using Equation (63), for $m=0,1,2,3,\cdots$ we get the following values

$$\begin{aligned} u_1(\Im,\Re,\tau) &= 4\sinh(\Im)\sinh(\Re)\frac{\tau^{\beta}}{(\beta)!}, \\ u_2(\Im,\Re,\tau) &= -8\frac{\beta(\beta-1)!\sinh(\Im)\sinh(\Re)\tau^{2\beta}}{(2\beta-1)!(\beta)!}, \\ u_3(\Im,\Re,\tau) &= 16\frac{\beta(2\beta-1)(\beta-1)!(2\beta-2)!\sinh(\Im)\sinh(\Re)\tau^{3\beta-2}}{(\beta)!(2\beta-1)!(3\beta-2)!}, \\ &\vdots \end{aligned} \tag{64}$$

Now using the values of $u_0,u_1,u_2,u_3,\cdots$, we get Shehu transformation solution for Example 4

$$\begin{aligned} u(\Im,\Re,\tau) &= \sinh(\Im)\sinh(\Re)-2\tau\sinh(\Im)\sinh(\Re)+4\frac{\sinh(\Im)\sinh(\Re)\tau^{\beta}}{(\beta)!}-8\frac{\beta(\beta-1)!\sinh(\Im)\sinh(\Re)\tau^{2\beta}}{(2\beta-1)!(\beta)!}+ \\ &\frac{16\beta(2\beta-1)(\beta-1)!(2\beta-2)!\sinh(\Im)\sinh(\Re)\tau^{3\beta-2}}{(\beta)!(2\beta-1)!(3\beta-2)!}+\cdots. \end{aligned} \tag{65}$$

After simplification, we get

$$u(\Im,\Re,\tau)=\sinh(\Im)\sinh(\Re)\left\{1-2\tau+4\frac{\tau^{\beta}}{\beta!}-8\frac{\beta(\beta-1)!\tau^{2\beta}}{(2\beta-1)!(\beta)!}+\frac{16\beta(2\beta-1)(\beta-1)!(2\beta-2)!\tau^{3\beta-2}}{(\beta)!(2\beta-1)!(3\beta-2)!}+\cdots\right\}. \tag{66}$$

For particular case $\beta\to 2$, the Shehu transform solution become as

$$u(\Im,\Re,\tau)=\sinh(\Im)\sinh(\Re)\left\{1-2\tau+4\frac{\tau^2}{2!}-8\frac{\tau^3}{3!}+16\frac{\tau^4}{4!}+\cdots\right\}. \tag{67}$$

The calculated result provide the exact solution in the close form

$$u(\Im,\Re,\tau)=\sinh(\Im)\sinh(\Re)e^{-2\tau}. \tag{68}$$

Figures 12 and 13, display the exact and analytical solutions of Example 4. The solution graph of SDM is very similarly to the exact solution of Example 4. In Figure 14, we plotted the solutions of Example 4 at different fractional-order β. The fractional-order solutions are found to be convergent towards the exact solution of Example 4. It is investigated from the solution analysis that the present method is a sophisticated technique to solve fractional-order problems.

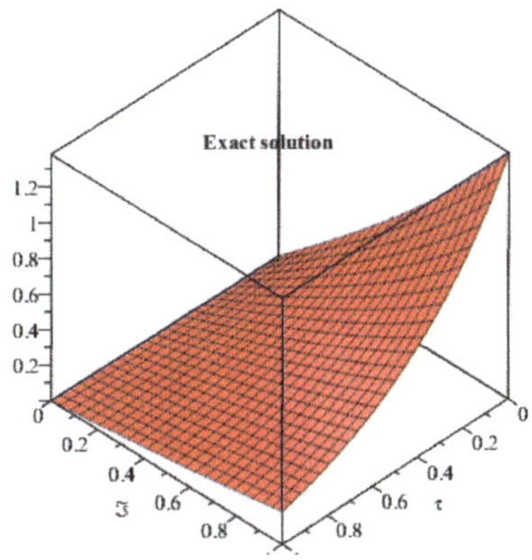

Figure 12. Exact solution of Example 4 at $\beta = 2$.

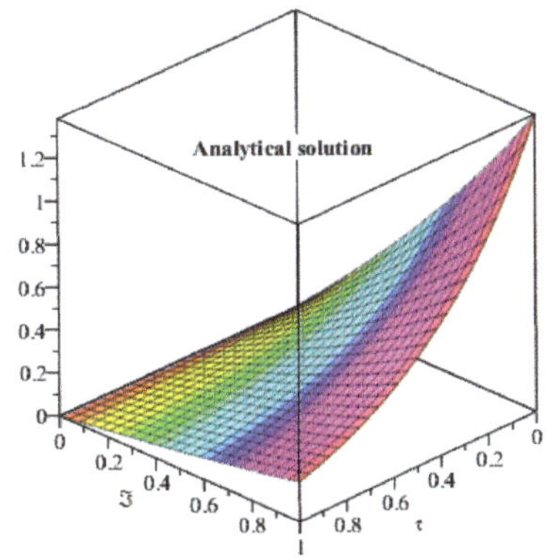

Figure 13. analytical solution of Example 4 at $\beta = 2$.

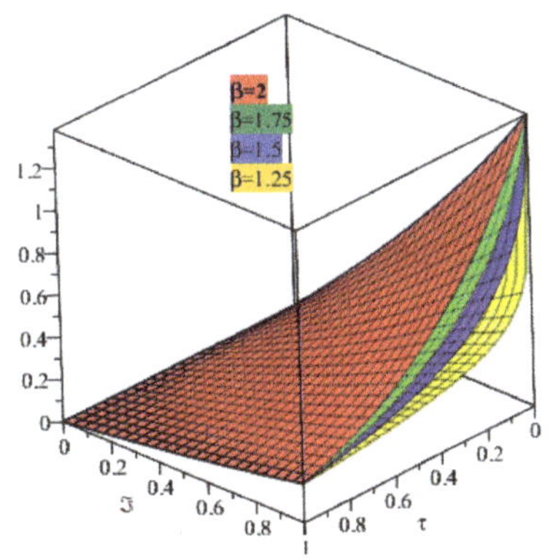

Figure 14. The solution graph at different fractional order β. of Example 4.

Example 5. *Consider the non-linear* $(2 + time fractional)$ *dimensional Burger's model:*

$$u_{\tau}^{\beta}(\Im, \Re, \tau) = u_{\Im\Im}(\Im, \Re, \tau) + u_{\Re\Re}(\Im, \Re, \tau) + u_{\Im}(\Im, \Re, \tau)u(\Im, \Re, \tau), \quad \beta \in (0, 1], \tag{69}$$

with initial condition

$$u(\Im, \Re, 0) = \Im + \Re. \tag{70}$$

If $\beta = 1$, then the exact solution of Equation (69) is

$$u(\Im, \Re, \tau) = \frac{\Im + \Re}{1 - \tau}. \tag{71}$$

Taking Shehu transform of Equation (69)

$$\frac{s^{\beta}}{\mu^{\beta}} \left\{ V(s, \mu) - \frac{\mu}{s} u(0) \right\} = S \left\{ u_{\Im\Im}(\Im, \Re, \tau) + u_{\Re\Re}(\Im, \Re, \tau) + u_{\Im}(\Im, \Re, \tau) u(\Im, \Re, \tau) \right\}, \tag{72}$$

The simplifying Equation (72), we get as

$$V(s, \mu) = \frac{\mu}{s} u(0) + \frac{\mu^{\beta}}{s^{\beta}} S \left\{ u_{\Im\Im}(\Im, \Re, \tau) + u_{\Re\Re}(\Im, \Re, \tau) + u_{\Im}(\Im, \Re, \tau) u(\Im, \Re, \tau) \right\}, \tag{73}$$

By applying inverse of Shehu transform, we get

$$u(\Im, \Re, \tau) = u(0) + \tau u'(0) + S^{-1} \left\{ \frac{\mu^{\beta}}{s^{\beta}} S \left\{ u_{\Im\Im}(\Im, \Re, \tau) + u_{\Re\Re}(\Im, \Re, \tau) + u_{\Im}(\Im, \Re, \tau) u(\Im, \Re, \tau) \right\} \right\}, \tag{74}$$

Thus we get the following recursive scheme

$$u_0(\Im, \Re, \tau) = u(0) = \Im + \Re, \tag{75}$$

$$u_{m+1}(\Im, \Re, \tau) = S^{-1} \left\{ \frac{\mu^{\beta}}{s^{\beta}} S \left\{ u_{m\Im\Im}(\Im, \Re, \tau) + u_{m\Re\Re}(\Im, \Re, \tau) + u_{m\Im}(\Im, \Re, \tau) u_m(\Im, \Re, \tau) \right\} \right\}. \tag{76}$$

For nonlinear term, use the Equation (12) in recursive scheme (76), we obtain

$$u_{m+1}(\Im, \Re, \tau) = S^{-1} \left\{ \frac{\mu^{\beta}}{s^{\beta}} S \left\{ u_{m\Im\Im}(\Im, \Re, \tau) + u_{m\Re\Re}(\Im, \Re, \tau) + \sum_{m=0}^{\infty} A_m(u_0, u_1, \cdots) \right\} \right\}. \tag{77}$$

Using Equation (77), for $m = 0, 1, 2, 3, \cdots$ we get the following values

$$\begin{aligned} u_1(\Im, \Re, \tau) &= (\Im + \Re) \frac{t^{\beta}}{(\beta)!}, \\ u_2(\Im, \Re, \tau) &= 2(\Im + \Re) \frac{\tau^{2\beta}}{(2\beta)!}, \\ u_3(\Im, \Re, \tau) &= 4(\Im + \Re) \frac{\tau^{3\beta}}{(3\beta)!} + (\Im + \Re)(2\beta)! \frac{t^{3\beta}}{\beta! \beta! (3\beta)!}, \\ &\vdots \end{aligned} \tag{78}$$

Now using the values of $u_0, u_1, u_2, u_3, \cdots$, we get Shehu transformation solution for example 5

$$u(\Im, \Re, \tau) = \Im + \Re + (\Im + \Re) \frac{\tau^{\beta}}{(\beta)!} + 2(\Im + \Re) \frac{\tau^{2\beta}}{(2\beta)!} + 4(\Im + \Re) \frac{\tau^{3\beta}}{(3\beta)!} + (\Im + \Re)(2\beta)! \frac{\tau^{3\beta}}{\beta! \beta! (3\beta)!} + \cdots. \tag{79}$$

After simplification, we get

$$u(\Im, \Re, \tau) = (\Im + \Re) \left\{ 1 + \frac{\tau^{\beta}}{(\beta)!} + 2 \frac{\tau^{2\beta}}{(2\beta)!} + 4 \frac{\tau^{3\beta}}{(3\beta)!} + (2\beta)! \frac{\tau^{3\beta}}{\beta! \beta! (3\beta)!} + \cdots \right\}. \tag{80}$$

For particular case $\beta \to 1$, the Shehu transform solution become as

$$u(\Im, \Re, \tau) = (\Im + \Re) \left\{ 1 + \tau + \tau^2 + \tau^3 + \cdots \right\}. \tag{81}$$

The calculated result provide the exact solution in the close form

$$u(\Im, \Re, \tau) = \frac{\Im + \Re}{1 - \tau}. \tag{82}$$

Figures 15 and 16 are plotted to discuss the exact and analytical solutions of Example 5. The SDM solutions are in good contact with the exact solution of the Example 5. Figures 17 and 18 are plotted to analyze the fractional-order solutions of Example 5 at fractional-order $\beta = 0.75$ and 0.50 respectively. The graphical analysis has verified the applicability of the proposed method.

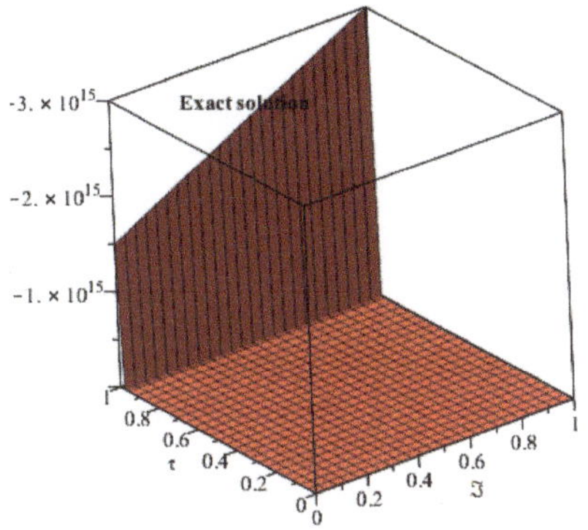

Figure 15. Exact solution of Example 5 at $\beta = 1$.

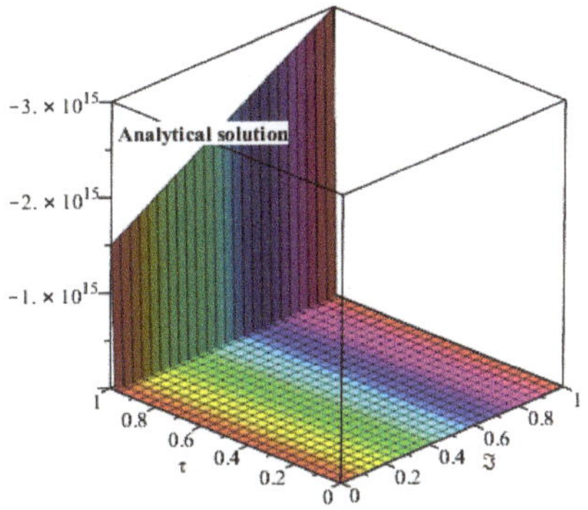

Figure 16. Represents the analytical solution of Example 5 at $\beta = 1$.

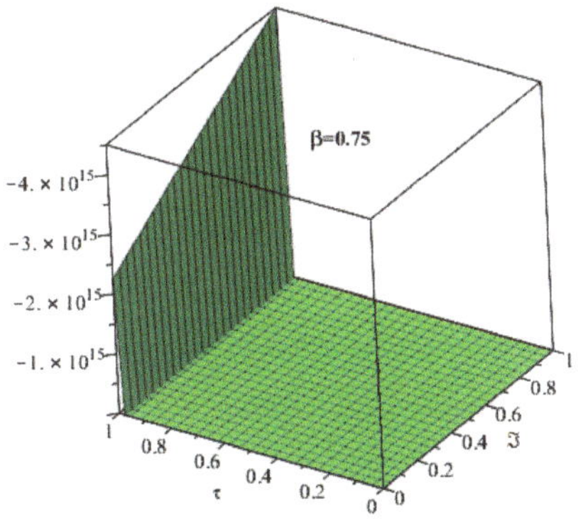

Figure 17. The solution of fractional-order $\beta = 0.75$ of Example 5.

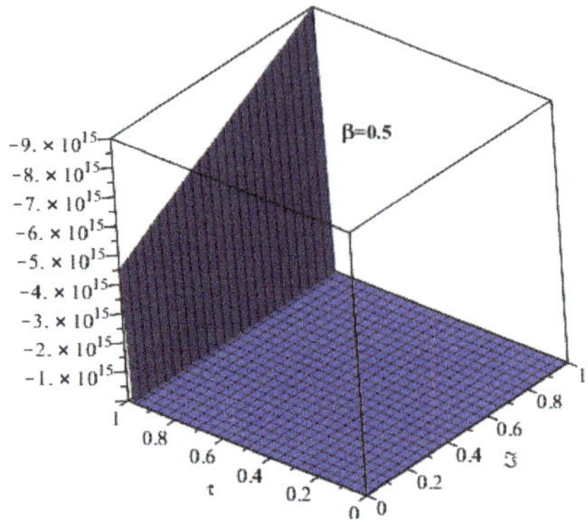

Figure 18. The solution of fractional-order $\beta = 0.5$ of Example 5.

5. Results and Discussion

In the present research work, we implemented a new analytical technique SDM for the solution of some important problems which are frequently arising in science and engineering, such as hyperbolic wave equation, heat equation, diffusion equation, telegraph and Burgers equations. The Caputo definition of fractional-derivative is used to define fractional-derivative. The proposed method is the combination of Shehu transformation and Adomian decomposition method which is known as Shehu decomposition method. For applicability and novelty of present method, we applied it different physical problems for applied sciences. These problems have been solved by using SDM for both fractional and integer-order of the targeted problems. In this connection some figure analysis have been done to demonstrate the obtained results in a sophisticated manner. It is investigated that SDM solution have a very close contact with the exact solution of the problems. It is also observed that the fractional-order problems are convergent towards the solution of an integer-order problem. Moreover, the high rate of convergence of the current method is noted during the simulation. It is calculated that the SDM can be considered as one of the best analytical technique to solve fractional partial differential equations.

6. Conclusions

In the present article, we presented some fractional-view analysis of physical problems, arising in science and engineering. A new and sophisticated analytical technique, which is known as Shehu transform decomposition method is implemented for both fractional and integer-orders of the problems. The Caputo definition of fractional derivative is used to express fractional-order derivative. For applicability and reliability of the proposed methods, some illustrative examples are presented from different areas of applied science. It has been investigated through graphical representation that the present technique provides an accurate and deserving analysis about the physical happening of the problems. It is observed through simulations of the present algorithm that as fractional-order of the derivative approaches to integer order of the problem then fractional-order solutions are convergent to integer-order solutions. Moreover, the present method is preferred as compared to other method because of its better rate of convergence. This direction motivates the researchers towards the implementation of the current method for other non-linear fractional partial differential equations.

Author Contributions: Conceptualization, R.S. and H.K.; Methodology, M.A.; Software, R.S.; Validation, R.S. and U.F.; Formal Analysis, R.S.; Investigation, R.S. and D.B.; Resources, H.K. and P.K.; Writing—Original Draft Preparation, R.S.; Writing—Review and Editing, H.K., U.F. and P.K.; Visualization, M.A.; Supervision, D.B., P.K.; Project Administration, P.K.; Funding Acquisition, P.K.

Funding: Center of Excellence in Theoretical and Computational Science (TaCS-CoE) Faculty of Science, King Mongkuts University of Technology Thonburi (KMUTT).

Conflicts of Interest: The authors declare no conflict of interest.

References

1. Abd-Elhameed, W.M.; Youssri, Y.H. Fifth-kind orthonormal Chebyshev polynomial solutions for fractional differential equations. *Comput. Appl. Math.* **2018**, *37*, 2897–2921. [CrossRef]
2. Shah, R.; Khan, H.; Farooq, U.; Baleanu, D.; Kumam, P.; Arif, M. A New Analytical Technique to Solve System of Fractional-Order Partial Differential Equations. *IEEE Access* **2019**, *7*, 150037–150050. [CrossRef]
3. Jain, S. Numerical analysis for the fractional diffusion and fractional Buckmaster equation by the two-step Laplace Adam-Bashforth method. *Eur Phys. J. Plus* **2018**, *133*, 19. [CrossRef]
4. Prakash, A. Analytical method for space-fractional telegraph equation by homotopy perturbation transform method. *Nonlinear Eng.* **2016**, *5*, 123–128. [CrossRef]
5. Khan, H.; Shah, R.; Baleanu, D.; Kumam, P.; Arif, M. Analytical Solution of Fractional-Order Hyperbolic Telegraph Equation, Using Natural Transform Decomposition Method. *Electronics* **2019**, *8*, 1015. [CrossRef]
6. Safari, M.; Ganji, D.D.; Moslemi, M. Application of He's variational iteration method and Adomian's decomposition method to the fractional KdV–Burgers–Kuramoto equation. *Comput. Math. Appl.* **2009**, *58*, 2091–2097. [CrossRef]
7. Kumar, D.; Tchier, F.; Singh, J.; Baleanu, D. An efficient computational technique for fractal vehicular traffic flow. *Entropy* **2018**, *20*, 259. [CrossRef]
8. Singh, J.; Kumar, D.; Baleanu, D.; Rathore, S. An efficient numerical algorithm for the fractional Drinfeld–Sokolov–Wilson equation. *Appl. Math. Comput.* **2018**, *335*, 12–24. [CrossRef]
9. Shivanian, E.; Jafarabadi, A. Time fractional modified anomalous sub-diffusion equation with a nonlinear source term through locally applied meshless radial point interpolation. *Mod. Phys. Lett. B* **2018**, *32*, 1850251. [CrossRef]
10. Ullah, S.; Khan, M.A.; Farooq, M. A new fractional model for the dynamics of the hepatitis B virus using the Caputo-Fabrizio derivative. *Eur. Phys. J. Plus* **2018**, *133*, 237. [CrossRef]
11. Qureshi, S.; Yusuf, A. Modeling chickenpox disease with fractional derivatives: From caputo to atangana-baleanu. *Chaos Solitons Fractals* **2019**, *122*, 111–118. [CrossRef]
12. Qureshi, S.; Yusuf, A.; Shaikh, A.A.; Inc, M.; Baleanu, D. Fractional modeling of blood ethanol concentration system with real data application. *Chaos Interdiscip. J. Nonlinear Sci.* **2019**, *29*, 013143. [CrossRef] [PubMed]
13. Khan, M.A.; Ullah, S.; Farooq, M. A new fractional model for tuberculosis with relapse via Atangana–Baleanu derivative. *Chaos Solitons Fractals* **2018**, *116*, 227–238. [CrossRef]
14. Jena, R.M.; Chakraverty, S. Residual Power Series Method for Solving Time-fractional Model of Vibration Equation of Large Membranes. *J. Appl. Comput. Mech.* **2019**, *5*, 603–615.
15. Jena, R.M.; Chakraverty, S. A new iterative method based solution for fractional Black–Scholes option pricing equations (BSOPE). *SN Appl. Sci.* **2019**, *1*, 95. [CrossRef]
16. Jena, R.M.; Chakraverty, S.; Jena, S.K. Dynamic response analysis of fractionally damped beams subjected to external loads using homotopy analysis method. *J. Appl. Comput. Mech.* **2019**, *5*, 355–366.
17. Abd-Elhameed, W.M.; Youssri, Y.H. Spectral tau algorithm for certain coupled system of fractional differential equations via generalized Fibonacci polynomial sequence. *Iran. J. Sci. Technol. Trans. A Sci.* **2019**, *43*, 543–554. [CrossRef]
18. Khan, H.; Shah, R.; Kumam, P.; Arif, M. Analytical Solutions of Fractional-Order Heat and Wave Equations by the Natural Transform Decomposition Method. *Entropy* **2019**, *21*, 597. [CrossRef]
19. Shah, R.; Khan, H.; Mustafa, S.; Kumam, P.; Arif, M. Analytical Solutions of Fractional-Order Diffusion Equations by Natural Transform Decomposition Method. *Entropy* **2019**, *21*, 557. [CrossRef]
20. Khan, M.A.; Ullah, S.; Okosun, K.O.; Shah, K. A fractional order pine wilt disease model with Caputo–Fabrizio derivative. *Adv. Differ. Equ.* **2018**, *2018*, 410. [CrossRef]
21. Singh, J.; Kumar, D.; Baleanu, D. On the analysis of fractional diabetes model with exponential law. *Adv. Differ. Equ.* **2018**, *2018*, 231. [CrossRef]
22. Qureshi, S.; Yusuf, A. Mathematical modeling for the impacts of deforestation on wildlife species using Caputo differential operator. *Chaos Solitons Fractals* **2019**, *126*, 32–40. [CrossRef]
23. Jumarie, G. Laplace's transform of fractional order via the Mittag–Leffler function and modified Riemann–Liouville derivative. *Appl. Math. Lett.* **2009**, *22*, 1659–1664. [CrossRef]

24. Kumar, S. A new analytical modelling for fractional telegraph equation via Laplace transform. *Appl. Math. Model.* **2014**, *38*, 3154–3163. [CrossRef]
25. Kazem, S. Exact solution of some linear fractional differential equations by Laplace transform. *Int. J. Nonlinear Sci.* **2013**, *16*, 3–11.
26. Namias, V. The fractional order Fourier transform and its application to quantum mechanics. IMA *J. Appl. Math.* **1980**, *25*, 241–265. [CrossRef]
27. Chen, C.M.; Liu, F.; Turner, I.; Anh, V. A Fourier method for the fractional diffusion equation describing sub-diffusion. *J. Comput. Phys.* **2007**, *227*, 886–897. [CrossRef]
28. Jiang, X.; Xu, M. The fractional finite Hankel transform and its applications in fractal space. *J. Phys. A Math. Theor.* **2009**, *42*, 385201. [CrossRef]
29. Gorenflo, R.; Iskenderov, A.; Luchko, Y. Mapping between solutions of fractional diffusion-wave equations. *Fract. Calculus Appl. Anal.* **2000**, *3*, 75–86.
30. Debnath, L.; Bhatta, D. *Integral Transforms and Their Applications*; Chapman and Hall/CRC: Boca Raton, FL, USA, 2014.
31. Yang, X.J. *Local Fractional Functional Analysis and Its Applications*; Asian Academic Publisher Limited: Hong Kong, China, 2011.
32. Neamaty, A.; Agheli, B.; Darzi, R. Applications of homotopy perturbation method and Elzaki transform for solving nonlinear partial differential equations of fractional order. *Theory Approx. Appl.* **2016**, *6*, 91–104.
33. Jena, R.M.; Chakraverty, S. Solving time-fractional Navier–Stokes equations using homotopy perturbation Elzaki transform. *SN Appl. Sci.* **2019**, *1*, 16. [CrossRef]
34. Taha, N.E.H.; Nuruddeen, R.I.; Abdelilah, K.; Hassan, S. Dualities between "Kamal and Mahgoub integral transforms" and "Some famous integral transforms". *Br. J. Appl. Sci. Technol.* **2017**, *20*, 1–8. [CrossRef]
35. Aboodh, K.S. The new integral transform "Aboodh Transform". *Glob. J. Pure Appl. Math.* **2013**, *9*, 35–43.
36. Aggarwal, S.; Chauhan, R. A comparative study of Mohand and Aboodh transforms. *Int. J. Res. Adv. Technol.* **2019**, *7*, 520–529. [CrossRef]
37. Kılıçman, A.; Gadain, H.E. On the applications of Laplace and Sumudu transforms. *J. Frankl. Inst.* **2010**, *347*, 848–862. [CrossRef]
38. Jena, R.M.; Chakraverty, S. Analytical solution of Bagley-Torvik equations using Sumudu transformation method. *SN Appl. Sci.* **2019**, *1*, 246. [CrossRef]
39. Mao, Z.; Shen, J. Hermite spectral methods for fractional PDEs in unbounded domains. *SIAM J. Sci. Comput.* **2017**, *39*, A1928–A1950. [CrossRef]
40. Maitama, S.; Zhao, W. New integral transform: Shehu transform a generalization of Sumudu and Laplace transform for solving differential equations. *arXiv* **2019**, arXiv:1904.11370.
41. Rudolf, H. (Ed.) *Applications of Fractional Calculus in Physics*; World Scientific: Singapore, 2000.
42. Wazwaz, A.M.; Gorguis, A. Exact solutions for heat-like and wave-like equations with variable coefficients. *Appl. Math. Comput.* **2004**, *149*, 15–29. [CrossRef]

applied sciences

MDPI

Article

Fractional View Analysis of Acoustic Wave Equations, Using Fractional-Order Differential Equations

Izaz Ali [1], Hassan Khan [1], Rasool Shah [1], Dumitru Baleanu [2,3], Poom Kumam [4,5,*] and Muhammad Arif [1]

1 Department of Mathematics, Abdul Wali Khan University, Mardan 23200, Pakistan; izaz.ali@awkum.edu.pk (I.A.); hassanmath@awkum.edu.pk (H.K.); rasoolshah@awkum.edu.pk (R.S.); marifmaths@awkum.edu.pk (M.A.)
2 Department of Mathematics, Faculty of Arts and Sciences, Cankaya University, 06530 Ankara, Turkey; dumitru@cankaya.edu.tr
3 Institute of Space Sciences, 077125 Magurele, Romania
4 Center of Excellence in Theoretical and Computational Science (TaCS-CoE) & Department of Mathematics, Faculty of Science, King Mongkut's University of Technology Thonburi (KMUTT), 126 Pracha Uthit Rd., Bang Mod, Thung Khru, Bangkok 10140, Thailand
5 Department of Medical Research, China Medical University Hospital, China Medical University, Taichung 40402, Taiwan
* Correspondence: poom.kum@kmutt.ac.th

Received: 1 November 2019; Accepted: 15 December 2019; Published: 15 January 2020

Abstract: In the present research work, a newly developed technique which is known as variational homotopy perturbation transform method is implemented to solve fractional-order acoustic wave equations. The basic idea behind the present research work is to extend the variational homotopy perturbation method to variational homotopy perturbation transform method. The proposed scheme has confirmed, that it is an accurate and straightforward technique to solve fractional-order partial differential equations. The validity of the method is verified with the help of some illustrative examples. The obtained solutions have shown close contact with the exact solutions. Furthermore, the highest degree of accuracy has been achieved by the suggested method. In fact, the present method can be considered as one of the best analytical techniques compared to other analytical techniques to solve non-linear fractional partial differential equations.

Keywords: homotopy perturbation method; variational iteration method; Laplace transform method; acoustic wave equations

1. Introduction

Recently, fractional calculus and fractional differential equations (FDEs) have attracted the attention of scientists, mathematicians and engineers. A number of important implementations have been evaluated in various fields of sciences and engineering, such as material engineering, viscoelastic, electrochemistry, electromagnetic and dynamics physics which are described by fractional partial differential equations (FPDEs) [1]. Analytical approaches to solve FDEs are of great interest. There is no technique which provides an exact solution to the FDEs. Approximate approaches must be obtained by using techniques of series solution or linearization [2], followed by the application of proper numerical discretization [3–5] and system solvers [6–8]. Non-linear phenomena appear in a number of fields of engineering and sciences, such as solid state physics, chemical kinetics, non-linear spectroscopy, fluid physics, computational biology, quantum mechanics and thermodynamics etc. The concept of non-linearity is designed by various higher-order nonlinear partial differential equations (PDEs). For all of the physical systems, fundamental phenomena are covered by their nonlinear concepts [9,10].

In this paper, Laplace Variational Homotopy Perturbation Method (LVHPM) is implemented to solve the following linear and non-linear fractional-order regularized long wave equations.

$$\frac{\partial^\beta v}{\partial \eta^\beta} + \frac{1}{2}\frac{\partial v^2}{\partial \xi} - \frac{\partial}{\partial \eta}\left(\frac{\partial^2 v}{\partial \xi^2}\right) = 0, \qquad 0 < \xi \leq 1, \qquad 0 < \beta \leq 1, \quad \eta > 0, \tag{1}$$

with initial condition

$$v(\xi, 0) = \xi,$$

$$\frac{\partial^\beta v}{\partial \eta^\beta} + \frac{\partial v}{\partial \xi} + v\frac{\partial v}{\partial \eta} - \frac{\partial}{\partial \eta}\left(\frac{\partial^2 v}{\partial \xi^2}\right) = 0, \qquad 0 < \xi \leq 1, \qquad 0 < \beta \leq 1, \quad \eta > 0, \tag{2}$$

with initial condition

$$v(\xi, 0) = 3\alpha \sec h^2(\beta\xi), \quad \alpha > 0, \quad \beta = \frac{1}{2}\sqrt{\frac{\alpha}{1+\alpha}}.$$

$$\frac{\partial^\beta v}{\partial \eta^\beta} + \frac{\partial v}{\partial \xi} - 2\frac{\partial}{\partial \eta}\left(\frac{\partial^2 v}{\partial \xi^2}\right) = 0, \qquad 0 < \xi \leq 1, \qquad 0 < \beta \leq 1, \quad \eta > 0, \tag{3}$$

with initial condition

$$v(\xi, 0) = e^{-\xi},$$

and

$$\frac{\partial^\beta v(\xi.\eta)}{\partial \eta^\beta} + \frac{\partial^4 v(\xi.\eta)}{\partial \xi^4} = 0, \qquad 0 < \xi \leq 1, \qquad 0 < \beta \leq 1, \quad \eta > 0, \tag{4}$$

with initial condition

$$v(\xi, 0) = \sin \xi. \tag{5}$$

Equation (1) is known as the fractional-order non-linear regularized long wave equation (RLWE); Equation (2) is known as the fractional-order non-linear general regularized long wave equation (GRLWE) and Equations (3) and (4) are known as fractional-order linear regularized long wave equations (RLWEs) [11].

The Benjamin Bona Mahony equation (BBME) also identified the regularized long wave (RLW) equation. This equation is the updated version of Korteweg–de Vries equation (KdV) for the modeling of tiny amplitude lengthy surface gravitational waves spreading unidirectionally in two dimensions. RLW equations have several implementations in certain areas of science, such as ion-acoustic waves in plasma, longitudinal dispersive waves in elastic rods, magneto-hydrodynamic waves in plasma, rotating tube flow and stress waves in compressed gas bubble mixes, etc. The RLW equations are described as useful models in applied physics and engineering for many significant physical structures. They also design many liquid flow nature issues where diffusion is significant, either in viscous or shock situations. It can be used to model any dissipation-related non-linear wave diffusion problem. Chemical reaction, heat conduction, mass diffusion, viscosity, thermal radiation or other sources may result from this dissipation, depending on problem modeling [12].

The RLW problem is a family of non-linear growth models that provides excellent designs for predicting natural phenomena. The algorithm was initially introduced to define the undular bore behavior [13]. It was also obtained from the research of acoustic plasma waves of water and ion. An analytical solution for the RLW equation was identified under restricted initial and boundary conditions in [14]. The fractional RLW equations also define numerous significant ocean science and engineering phenomena such as long-wave and small frequency shallow water waves. The non-linear waves modeled on the fractional equations of RLW are of significant interest for several scientists in ocean shallow waves of liquid. The mathematically modeled non-linear waves in the ocean were the fractional RLW equations. Indeed, huge surface waves identified as the tsunami are described as fractional RLW equations. The huge internal waves in the interior of the ocean, resulting from the difference in temperature, that may destruct marine ships could be defined as fractional RLW equations in the current, highly efficient method.

In recent decades, many researchers and scientists have used analytical methods to solve these types of problems such as homotopy perturbation Sumudu transform method (HPSTM) [11], Adomian decomposition method (ADM) [15,16], least-squares method [17], optimal homotopy perturbation method [18], variational iteration method (VIM) homotopy perturbation method (HPM) [19] and He's homotopy perturbation method [20]. It is observed that these methods have certain deficiencies like calculation of Adomian polynomials, determination of Lagrange multiplier, divergent results and a huge volume of calculations. As a result, a modified analytical technique which is known as VHPTM was introduced to solve differential equations of fractional-order. VHPTM is the combination of three well-known techniques namely, homotopy perturbation method, Laplace transform and variational iteration method. The present method uses the Lagrange multiplier that can limit the consecutive implementation of integral operator and unmanageable computational cost. It is still maintaining higher degree of accuracy. VHPTM [21–24] has an excellent scheme and absorbs all the beneficial characteristics of VIM and HPM.

Finally, He's polynomials have been used in the correction fractional formula to develop the homotopy perturbation method. It is observed that the proposed method is implemented without any use of transformation, discretization and it was found to be free from the generating round off error. Usually, the method of variable separable needs both initial and boundary points to operate, but the present method provides an analytical solution by using initial conditions only. There is a clear advantage of the suggested method that it works without any use of Adomian polynomials, as required by the Adomian decomposition method. Results of the analysis show that the suggested method produces the solution in a series of fast convergence that can result in a closed solution [25–30].

2. Preliminaries Concepts

Definition 1. *Laplace transformation of $\rho(\eta)$, $\eta > 0$ represented as [13]*

$$Q(s) = \mathcal{L}[\rho(\eta)] = \int_0^\infty e^{-s\eta}\rho(\eta)d\eta.$$

Theorem 1. *The convolution of Laplace transform is*

$$\mathcal{L}[\rho_1 \times \rho_2] = \mathcal{L}[\rho_1(\eta)] \times \mathcal{L}[\rho_2(\eta)],$$

here $\rho_1 \times \rho_2$, define the convolution between ρ_1 and ρ_2 ,

$$(\rho_1 \times \rho_2)\eta = \int_0^\tau \rho_1(\tau)\rho_2(\eta - \tau)d\eta.$$

Laplace transform of fractional derivative

$$\mathcal{L}\left(D_\eta^\beta \rho(\eta)\right) = s^\beta Q(s) - \sum_{k=0}^{n-1} s^{\beta-1-k}\rho^{(k)}(0), \quad n-1 < \beta < n.$$

where $Q(s)$ is the Laplace transformation of $\rho(\eta)$.

Definition 2. *The Riemann-Liouville fractional integral operator of order $\beta \geq 0$ of a function $f \in C_\mu$, $\mu \geq -1$ is described in [21]*

$$I_\xi^\beta g(\xi) = \frac{1}{\Gamma(\beta)}\int_0^\xi (\xi - s)^{\beta-1} g(s)ds,$$

where Γ *represent the gamma function as,*

$$\Gamma(\beta) = \int_0^\infty e^{-\xi}\xi^{\beta-1}d\xi \qquad \beta \in \mathbb{C}.$$

Definition 3. *The fractional derivative of* $g(\eta)$ *in the Caputo sense is defined as*

$$D^\beta g(\eta) = \frac{\partial^\beta g(\eta)}{\partial \eta^\beta} = \begin{cases} I^{m-\beta}\left[\frac{\partial^m g(\eta)}{\partial \eta^m}\right], & \text{if } m-1 < \beta < m, \quad m \in \mathbb{N} \\ \frac{\partial^m g(\eta)}{\partial \eta^m}, & \beta = m. \end{cases}$$

Lemma 1. *If* $\tilde{m} - 1 < \beta \leq \tilde{m}$ *with* $\tilde{m} \in \mathbb{N}$ *and* $g \in \mathbb{C}_\eta$ *with* $\eta \geq -1$, *then*

$$I^\beta I^a g(\xi) = I^{\beta+a} g(\xi), \qquad a, \beta \geq 0.$$
$$I^\beta \xi^\lambda = \frac{\Gamma(\lambda+1)}{\Gamma(\gamma+\lambda+1)}\xi^{\beta+\lambda}, \qquad \beta > 0, \lambda > -1, \quad \xi > 0.$$
$$I^\beta D^\beta g(\xi) = g(\xi) - \sum_{k=0}^{\tilde{m}-1} g^{(k)}(0^+)\frac{\xi^k}{k!}, \quad \text{for} \quad \xi > 0, \tilde{m} - 1 < \beta \leq \tilde{m}.$$

Definition 4. *Function of Mittag-Leffler,* $E_{\alpha,\beta}(\eta)$ *for* $\alpha, \beta > 0$ *is defined as*

$$E_{\alpha,\beta}(\eta) = \sum_{k=0}^\infty \frac{\eta^k}{\Gamma(k\alpha+\beta)}, \qquad \alpha, \beta > 0, \quad \eta \in \mathbb{C}.$$

3. The Procedure of VHPTM

To demonstrate, the fundamental concept of the present method [21,22], we are considering

$$D_\eta^\beta v(\xi,\eta) + \bar{R}v(\xi,\eta) + \bar{N}v(\xi,\eta) = f(\xi,\eta), \tag{6}$$

with initial condition

$$v(\xi, 0) = g(\xi),$$

where $f(\xi,\eta)$ is an inhomogeneous term, $\bar{R}$ and $\bar{N}$ are particular linear and non-linear differential operators and $D_\eta^\beta v(\xi,\eta)$ is the Caputo fractional derivative of $v(\xi,\eta)$.

By taking Laplace transform of Equation (6) on both sides, we get

$$\pounds_\eta\{v(\xi,\eta)\} - \sum_{k=0}^{m-1} s^{\beta-1-k}\frac{\partial^k v(\xi,\eta)}{\partial^k \eta}|_{t=0} = -\pounds\left\{\bar{R}v(\xi,\eta) + \bar{N}v(\xi,\eta) - f(\xi,\eta)\right\},$$

where $v(s) = \pounds_\eta(v(\xi,\eta)) = \int_0^\infty e^{-s\eta}v(\eta)d\eta$.

We can build a functional correction according to the variation iteration method

$$\pounds_\eta\{v_{j+1}(\xi,\eta)\} = \pounds\{v_j(\xi,\eta)\} + \lambda(s)\left[s^\beta \pounds\left\{v_\eta(\xi,\eta) - \sum_{k=0}^{m-1} s^{\beta-1-k}\frac{\partial^k v(\xi,\eta)}{\partial^k \eta}|_{t=0} + \pounds\left\{\bar{R}v(\xi,\eta) + \bar{N}v(\xi,\eta) - f(\xi,\eta)\right\}\right\}\right], \tag{7}$$

where $\lambda(s)$ is the Lagrange multiplier. Here we put $\lambda(s) = \frac{-1}{s^\beta}$ [22].

Applying inverse Laplace of Equation (7)

$$v_{j+1}(\xi,\eta) = v_j(\xi,\eta) - \pounds^{-1}\left[\frac{1}{s^\beta}\pounds\left\{s^\beta\frac{\partial v}{\partial \eta} + \bar{R}v_j(\xi,\eta) + \bar{N}v_j(\xi,\eta) - f(\xi,\eta)\right\}\right]. \tag{8}$$

The basic idea in the procedure of homotopy perturbation method is that the solution can be written as a series in powers of p:

$$v(\xi,\eta)=\sum_{j=0}^{\infty}p^{j}v_{j}(\xi,\eta)=v_{0}+pv_{1}+p^{2}v_{2}+p^{3}v_{3}+\cdots, \tag{9}$$

where the non-linear expression can be expressed as

$$\bar{N}v(\xi,\eta)=\sum_{j=0}^{\infty}p^{j}\bar{H}_{j}(v). \tag{10}$$

$\bar{H}_j$ is He's polynomials,

$$\bar{H}_{j}(v_{0}+v_{1}+\cdots+u_{j})=\frac{1}{j!}\frac{\partial^{j}}{\partial p^{j}}\left[\bar{N}\left(\sum_{i=0}^{\infty}p^{i}v_{i}\right)\right]. \tag{11}$$

The technique of fractional VHPTM of Equation (8) with He's polynomials.

$$\sum_{j=0}^{\infty}p^{j}v_{j}(\xi,\eta)=\sum_{j=0}^{\infty}p^{j}v_{j}(\xi,\eta)+\pounds^{-1}\left[\lambda(s)\pounds\left\{\sum_{j=0}^{\infty}p^{j}\frac{\partial^{\beta}v_{j}}{\partial\eta^{\beta}}(\xi,s)+\sum_{j=0}^{\infty}p^{j}\bar{R}v_{j}(\xi,\eta)+\sum_{j=0}^{\infty}p^{j}\bar{H}_{j}(v)-f(\xi,\eta)\right\}\right]. \tag{12}$$

By comparing the coefficients of like power of p on both sides of Equation (12), we get the VHPTM solution of the given problem.

Theorem 2. *Let ξ and $\mathcal{Y}$ be two Banach spaces and $T:\xi\rightarrow\mathcal{Y}$ be a contractive nonlinear operator, such that for all $v;v^{*}\in;\xi,||T(v)-T(v^{*})||\leq K||v-v^{*}||,\quad 0<K<1\,y$ [31]. Then, in view of Banach contraction theorem, T has a unique fixed point v, such that $Tv=v$: Let us write the generated series (12), by the Laplace decomposition method as*

$$\xi_{m}=T(\xi_{m-1}),\quad \xi_{m-1}=\sum_{m=1}^{m-1}v_{j},\quad m=0,1,2,\cdots$$

and supposed that $\xi_{0}=v_{0}\in\mathcal{S}_{p}(v)$, where $\mathcal{S}_{p}(v)=\{v^{*}\in\xi:||v-v^{*}||<p\}$ then, we have

$$(B_{1})\xi_{m}\in\mathcal{S}_{p}(v)$$
$$(B_{2})\lim_{m\rightarrow\infty}\xi_{m}=v.$$

Proof. (B_1) In view of mathematical induction for $m=1$, we have

$$||\xi_{1}-v_{1}||=||T(\xi_{0}-T(v))||\leq K||v_{0}-v||.$$

Let the result be true for $m-1$, then

$$||\xi_{m-1}-v||\leq K^{m-1}||v_{0}-v||.$$

We have

$$||\xi_{m}-v||=||T(\xi_{m-1}-T(v))||\leq K||\xi_{m-1}-v||\leq K^{m}||v_{0}-v||.$$

Hence, using (B_1), we have

$$||\xi_{m}-v||\leq K^{m}||v_{0}-v||\leq K^{m}p<p,$$

which implies that $\xi_m\in\mathcal{S}_p(v)$.

(B_2): Since $||\xi_m - v|| \leq K^m||v_0 - v||$ and as a $\lim_{m\to\infty} K^m = 0$.
Therefore; we have $\lim_{m\to\infty} ||\xi_m - v|| = 0 \Rightarrow \lim_{m\to\infty} \xi_m = v$. □

4. Numerical Examples

4.1. Example

We consider time fractional-order non-linear RLW equation

$$\frac{\partial^\beta v}{\partial \eta^\beta} + \frac{1}{2}\frac{\partial v^2}{\partial \xi} - \frac{\partial \partial^2 v}{\partial \eta \partial \xi^2} = 0, \qquad 0 < \xi \leq 1, \qquad 0 < \beta \leq 1, \quad \eta > 0, \tag{13}$$

initial condition is

$$v(\xi, 0) = \xi, \tag{14}$$

By using Equation (12), the fractional PDE given in Equation (13) can be written as

$$\sum_{j=0}^{\infty} p^j v_{j+1}(\xi,\eta) = \sum_{j=0}^{\infty} p^j v_j(\xi,\eta) + \pounds^{-1}\left[\lambda(s)\pounds\left\{s^\beta \frac{\partial v_j(\xi,\eta)}{\partial \eta} + \frac{1}{2}\frac{\partial v_j^2(\xi,\eta)}{\partial \xi} - \frac{\partial \partial^2 v_j(\xi,\eta)}{\partial \eta \partial \xi^2}\right\}\right], \tag{15}$$

where $\lambda(s)$ is the Lagrange multiplier

$$\lambda(s) = \frac{-1}{s^\beta}.$$

Applying VHPTM using He's polynomials,

$$\begin{aligned}\sum_{j=0}^{\infty} p^j v_{j+1}(\xi,\eta) &= \sum_{j=0}^{\infty} p^j v_j(\xi,\eta) - \sum_{j=0}^{\infty} p^j \pounds^{-1}\left[\frac{1}{s^\beta}\pounds\left\{s^\beta\left(\frac{\partial v_0}{\partial \eta} + p\frac{\partial v_1}{\partial \eta} + p^2\frac{\partial v_2}{\partial \eta} + \cdots\right)\right.\right.\\ &+\frac{1}{2}\frac{\partial}{\partial \xi}\left\{v_0^2 + p(2v_0v_1) + p^2(2v_0v_2 + v_1^2) + \cdots\right\} - \left.\left.\left\{p^0\frac{\partial \partial^2 v_0}{\partial \eta \partial \xi^2} + p^1\frac{\partial \partial^2 v_1}{\partial \eta \partial \xi^2} + p^2\frac{\partial \partial^2 v_2}{\partial \eta \partial \xi^2} + \cdots\right\}\right\}\right],\end{aligned} \tag{16}$$

Comparing the coefficients of p

$$\begin{aligned}
&v_0(\xi,\eta) = \xi,\\
&p^1 v_1(\xi,\eta) = p^1 v_0(\xi,\eta) - p^1\pounds^{-1}\left[\frac{1}{s^\beta}\pounds\left\{s^\beta\frac{\partial v_0(\xi,\eta)}{\partial \eta} + \frac{1}{2}\frac{\partial}{\partial \xi}v_0^2(\xi,\eta) - \frac{\partial \partial^2 v_0(\xi,\eta)}{\partial \eta \partial \xi^2}\right\}\right],\\
&v_1(\xi,\eta) = \xi - \xi\frac{\eta^\beta}{\Gamma(\beta+1)},\\
&p^2 v_2(\xi,\eta) = p^2 v_1(\xi,\eta) - p^2\pounds^{-1}\left[\frac{1}{s^\beta}\pounds\left\{s^\beta\frac{\partial v_1(\xi,\eta)}{\partial \eta} + \frac{1}{2}\frac{\partial}{\partial \xi}(2v_0v_1) - \frac{\partial \partial^2 v_1}{\partial \eta \partial \xi^2}\right\}\right],\\
&v_2(\xi,\eta) = \xi - \xi\frac{\eta^\beta}{\Gamma(\beta+1)} + 2\xi\frac{\eta^{2\beta}}{\Gamma(2\beta+1)},\\
&p^3 v_3(\xi,\eta) = p^3 v_2(\xi,\eta) - p^3\pounds^{-1}\left[\frac{1}{s^\beta}\pounds\left\{s^\beta\frac{\partial v_1}{\partial \eta} + \frac{1}{2}(2v_0v_2 + v_1^2) - \frac{\partial \partial^2 v_2}{\partial \eta \partial \xi^2}\right\}\right],\\
&v_3(\xi,\eta) = \xi - \xi\frac{\eta^\beta}{\Gamma(\beta+1)} + 2\xi\frac{\eta^{2\beta}}{\Gamma(2\beta+1)} - \xi\frac{\Gamma(2\gamma+1)\eta^{3\beta}}{(\Gamma(2\gamma+1))^2\Gamma(3\beta+1)} - 4\xi\frac{\eta^{3\beta}}{\Gamma(3\beta+1)},\\
&\vdots
\end{aligned}$$

The analytical expression is therefore obtained in the following way

$$v(\xi,\eta)=\xi-\xi\frac{\eta^{\beta}}{\Gamma(\beta+1)}+2\xi\frac{\eta^{2\beta}}{\Gamma(2\beta+1)}-\xi\frac{\Gamma(2\gamma+1)\eta^{3\beta}}{(\Gamma(\gamma+1))^{2}\Gamma(3\beta+1)}-4\xi\frac{\eta^{3\beta}}{\Gamma(3\beta+1)}+\cdots. \tag{17}$$

If $\beta=1$ the series form is

$$v(\xi,\eta)=\xi\left(1-\eta+\eta^{2}-\eta^{3}+\cdots\right). \tag{18}$$

The exact solution at $\beta=1$

$$v(\xi,\eta)=\frac{\xi}{1+\eta}. \tag{19}$$

Figure 1, shows the plot of exact and VHPTM solutions of example 4.1, at integer-order $\beta=1$. It is confirmed from the figure that both exact and VHPTM solutions are in good contact with each other. In Figure 2, the VHPTM solutions at different fractional-orders $\beta=1,0.8,0.6$ and 0.4 are calculated. Investigations show that the solutions at different fractional-orders are convergent to the solution of an integer problem as the fractional-order approaches to an integer-order.

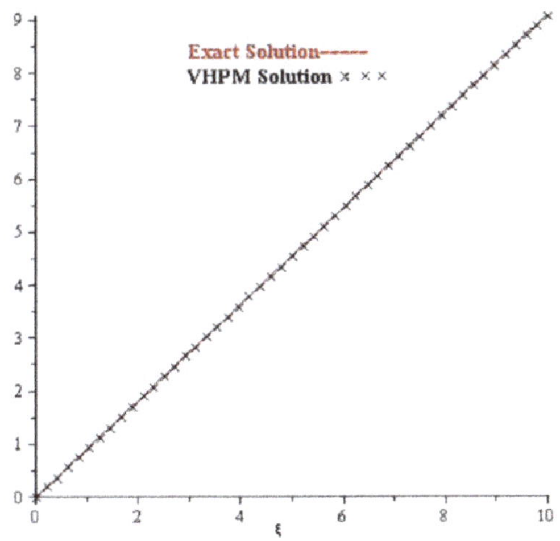

Figure 1. Variational homotopy perturbation transform method solution of example 4.1 for $\beta=1$.

In Table 1, we compared the solutions of VHPTM and VIM at an integer-order $\beta=1$ for example 4.1. In addition, the solutions at fractional-orders $\beta=0.55$ and $\beta=0.75$ are listed in the table. It is observed that VHPTM solutions are almost identical with each other. The results given in the table support the applicability of the VHPTM.

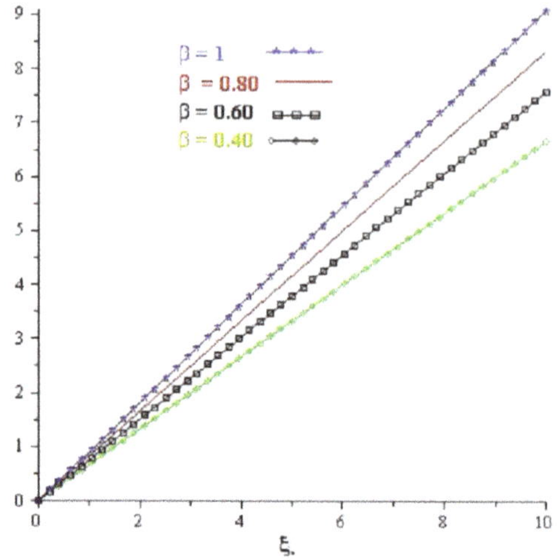

Figure 2. Variational homotopy perturbation transform method solution of example 4.1 at different fractional order $\beta = 1, 0.8, 0.6, 0.4$.

Table 1. Comparison of VHPTM and variational iteration method (VIM) [19] of example 1 at $\eta = 0.1$.

	VHPTM	VHPTM	VHPTM	Exact	Absolute Error	Absolute Error
ξ	$\beta = 0.55$	$\beta = 0.75$	$\beta = 1$		$VIM(\beta = 1)$	$VHPTM(\beta = 1)$
0.1	0.0966295721	0.0993928867	0.0999000999	0.0999001000	1×10^{-10}	1×10^{-10}
0.2	0.1932591443	0.1987857735	0.1998001998	0.1998002000	2×10^{-10}	2×10^{-10}
0.3	0.2898887165	0.2981786602	0.2997002997	0.2997003000	3×10^{-10}	3×10^{-10}
0.4	0.3865182887	0.3975715470	0.3996003996	0.3996004000	4×10^{-10}	4×10^{-10}
0.5	0.4831478608	0.4969644337	0.4995004995	0.4995005000	5×10^{-10}	5×10^{-10}
0.6	0.5797774330	0.5963573204	0.5994005994	0.5994006000	6×10^{-10}	6×10^{-10}
0.7	0.6764070052	0.6957502072	0.6993006993	0.6993007000	7×10^{-10}	7×10^{-10}
0.8	0.7730365774	0.7951430939	0.7992007992	0.7992008000	8×10^{-10}	8×10^{-10}
0.9	0.8696661495	0.8945359807	0.8991008991	0.8991009000	9×10^{-10}	9×10^{-10}
1	0.9662957217	0.9939288674	0.9990009990	0.9990010000	10×10^{-09}	10×10^{-09}

4.2. Example

We consider time fractional-order non-linear GRLW equation

$$\frac{\partial^\beta v}{\partial \eta^\beta} + \frac{\partial v}{\partial \xi} + v\frac{\partial v}{\partial \xi} - \frac{\partial \partial^2 v}{\partial \eta \partial \xi^2} = 0, \qquad 0 < \xi \leq 1, \qquad 0 < \beta \leq 1, \quad \eta > 0, \tag{20}$$

with initial condition

$$v(\xi, 0) = 3\alpha \sec h^2(\beta\xi), \quad \alpha > 0, \quad \beta = \frac{1}{2}\sqrt{\frac{\alpha}{1+\alpha}}. \tag{21}$$

By using Equation (12), the fractional PDE given in Equation (20) can be written as

$$\sum_{j=0}^{\infty} p^j v_{j+1}(\xi,\eta) = \sum_{j=0}^{\infty} p^j v_j(\xi,\eta) + p^j \pounds^{-1}\left[\lambda(s)\pounds\left\{s^\beta \frac{\partial v_j(\xi,\eta)}{\partial \eta} + \frac{\partial v_j(\xi,\eta)}{\partial \xi} + v_j(\xi,\eta)\frac{\partial v_j(\xi,\eta)}{\partial \xi} - \frac{\partial \partial^2 v_j(\xi,\eta)}{\partial \eta \partial \xi^2}\right\}\right], \tag{22}$$

where $\lambda(s)$ is the Lagrange multiplier

$$\lambda(s) = \frac{-1}{s^\beta}.$$

Applying VHPTM using He's polynomials,

$$\sum_{j=0}^{\infty} p^j v_j(\xi,\eta) = \sum_{j=0}^{\infty} p^j v_j(\xi,\eta) - p\pounds^{-1}\left[\frac{1}{s^\beta}\pounds\left\{s^\beta\left(\frac{\partial v_0}{\partial \eta} + p\frac{\partial v_1}{\partial \eta} + p^2\frac{\partial v_1}{\partial \eta} + \cdots\right) + \left\{\frac{\partial v_0}{\partial \xi} + p^1\frac{\partial v_1}{\partial \xi} + p^2\frac{\partial v_2}{\partial \xi} + \cdots\right\}\right.\right.$$
$$\left.\left. + \left\{v_0\frac{\partial v_0}{\partial \xi} + p(v_0\frac{\partial v_1}{\partial \xi} + v_1\frac{\partial v_0}{\partial \xi}) + p^2(v_0\frac{\partial v_2}{\partial \xi} + v_1\frac{\partial v_1}{\partial \xi} + v_2\frac{\partial v_0}{\partial \xi}) + \cdots\right\} - \left\{\frac{\partial\partial^2 v_0}{\partial\eta\partial\xi^2} + p^1\frac{\partial\partial^2 v_1}{\partial\eta\partial\xi^2} + p^2\frac{\partial\partial^2 v_2}{\partial\eta\partial\xi^2} + \cdots\right\}\right\}\right]. \quad (23)$$

Comparing the coefficients of p

$$v_0(\xi,\eta) = 3\alpha\sec h^2(\beta\xi),$$
$$p^1 v_1(\xi,\eta) = p^1 v_0(\xi,\eta) - p^1\pounds^{-1}\left[\frac{1}{s^\beta}\pounds\left\{s^\beta\frac{\partial v_0}{\partial \eta} + \frac{\partial v_0}{\partial \xi} + v_0\frac{\partial v_0}{\partial \xi} - \frac{\partial\partial^2 v_0}{\partial\eta\partial\xi^2}\right\}\right],$$
$$v_1(\xi,\eta) = 3\alpha\sec h^2(\beta\xi) + 3\alpha\beta\{1 + 6\alpha\beta + \cosh(2\beta\xi)\}\sec h^4(\beta\xi)\tanh(\beta\xi)\frac{\eta^\beta}{\Gamma(\beta+1)},$$
$$p^2 v_2(\xi,\eta) = p^2 v_1(\xi,\eta) - p^2\pounds^{-1}\left[\frac{1}{s^\beta}\pounds\left\{s^\beta\frac{\partial v_1}{\partial \eta} + \frac{\partial v_1}{\partial \xi} + v_0\frac{\partial v_1}{\partial \xi} + v_1\frac{\partial v_0}{\partial \xi} - \frac{\partial\partial^2 v_1}{\partial\eta\partial\xi^2}\right\}\right],$$
$$v_2(\xi,\eta) = 3\alpha\sec h^2(\beta\xi) + 3\alpha\beta\{1 + 6\alpha\beta + \cosh(2\beta\xi)\}\sec h^4(\beta\xi)\tanh(\beta\xi)\frac{\eta^\beta}{\Gamma(\beta+1)}$$
$$-\frac{3}{32}\alpha\beta^2\{-8 - 96\alpha - 576\alpha^2 + 3(-3 - 16\alpha + 144\alpha^2)\cosh(2\beta\xi) + 48\alpha\cosh(4\beta\xi) + \cosh(6\beta\xi)\}$$
$$\sec h^8(\beta\xi)\frac{\eta^{2\beta}}{\Gamma(2\beta+1)},$$

.

.

.

The analytical expression is therefore obtained in the following way

$$v(\xi,\eta) = 3\alpha\sec h^2(\beta\xi) + 3\alpha\beta\{1 + 6\alpha\beta + \cosh(2\beta\xi)\}\sec h^4(\beta\xi)\tanh(\beta\xi)\frac{\eta^\beta}{\Gamma(\beta+1)}$$
$$-\frac{3}{32}\alpha\beta^2\{-8 - 96\alpha - 576\alpha^2 + 3(-3 - 16\alpha + 144\alpha^2)\cosh(2\beta\xi) + 48\alpha\cosh(4\beta\xi) + \cosh(6\beta\xi)\}$$
$$\sec h^8(\beta\xi)\frac{\eta^{2\beta}}{\Gamma(2\beta+1)} + \frac{1}{32}\alpha\beta^3\{-85 - 1416\alpha - 8496\beta^2 - 2937\beta^3 + 4(-31 - 432\alpha - 1584\alpha^2 + 3456\alpha^3) \quad (24)$$
$$\cosh(2\beta\xi) - 4(11 + 54\alpha - 540\alpha^2)\cosh(4\beta\xi) - 4\cosh(6\beta\xi) + 96\alpha\cosh(6\beta\xi) + \cosh(8\beta\xi) + \sec h^8(\beta\xi)$$
$$\tanh(\beta\xi)\}\frac{\eta^{3\beta}}{\Gamma(3\beta+1)} + \cdots$$

The exact solution at $\beta = 1$

$$v(\xi,\eta) = 3\alpha\sec h^2(\beta(\xi - (1+\alpha)\eta)). \quad (25)$$

In Figure 3, we compared the analytical solution of VHPTM with the exact solution of example 4.2. The comparison has shown the close contact between VHPTM solution and exact solution of the problems. Figure 4, represents VHPTM solution at different fractional-orders $\beta = 1, 0.8, 0.6$ and 0.4 The convergence analysis of fractional-order problems are convergent towards the integer-order problem of example 4.2, as observed.

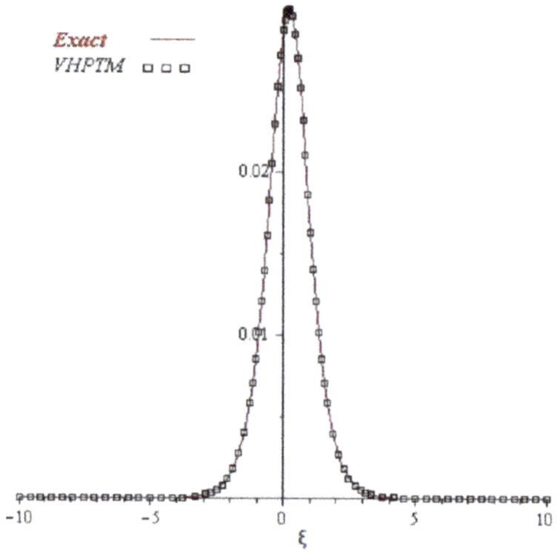

Figure 3. Variational homotopy perturbation transform method solution of example 4.2 for $\beta = 1$.

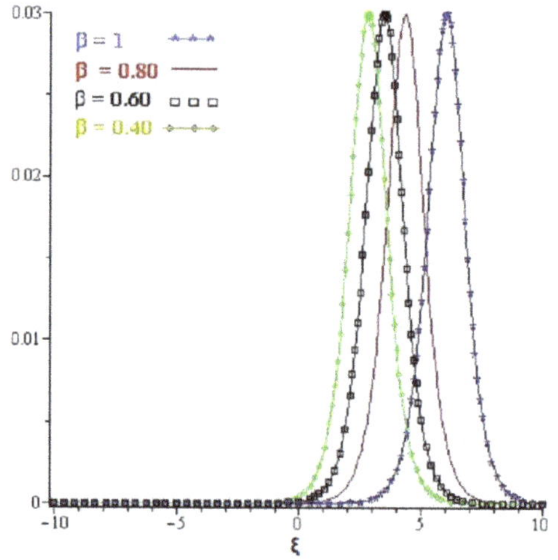

Figure 4. Variational homotopy perturbation transform method solution of example 4.2 at different fractional order $\beta = 1, 0.8, 0.6, 0.4$.

4.3. Example

We consider time fractional-order linear RLW equation

$$\frac{\partial^\beta v}{\partial \eta^\beta} + \frac{\partial v}{\partial \xi} - 2\frac{\partial \partial^2 v}{\partial \eta \partial \xi^2} = 0, \qquad 0 < \xi \leq 1, \qquad 0 < \beta \leq 1, \quad \eta > 0, \tag{26}$$

initial condition is

$$v(\xi, 0) = e^{-\xi}, \tag{27}$$

By using Equation (12), the fractional PDE given in Equation (26) can be written as

$$\sum_{j=0}^{\infty} p^j v_{j+1}(\xi,\eta) = \sum_{j=0}^{\infty} p^j v_j(\xi,\eta) + p^j \pounds^{-1}\left[\lambda(s)\pounds\left\{s^\beta \frac{\partial^\beta v_j}{\partial\eta^\beta}(\xi,\eta) + \frac{\partial v_j}{\partial\xi} - 2\frac{\partial\partial^2 v_j}{\partial\eta\partial\xi^2}\right\}\right], \tag{28}$$

where $\lambda(s)$ is the Lagrange multiplier

$$\lambda(s) = \frac{-1}{s^\beta}.$$

Applying VHPTM using He's polynomials,

$$\sum_{j=0}^{\infty} p^j v_j(\xi,\eta) = \sum_{j=0}^{\infty} p^j v_j(\xi,\eta) - p\pounds^{-1}\left[\frac{1}{s^\beta}\pounds\left\{s^\beta\left(\frac{\partial v_0}{\partial\eta} + p\frac{\partial v_1}{\partial\eta} + p^2\frac{\partial v_2}{\partial\eta} + \cdots\right)\right.\right.$$
$$\left.\left. + \frac{\partial}{\partial\xi}\left\{v_0 + pv_1 + p^2 v_2 + \cdots\right\} - 2\left\{\frac{\partial\partial^2 v_0}{\partial\eta\partial\xi^2} + p^1\frac{\partial\partial^2 v_1}{\partial\eta\partial\xi^2} + p^2\frac{\partial\partial^2 v_2}{\partial\eta\partial\xi^2} + \cdots\right\}\right\}\right]. \tag{29}$$

Comparing the coefficients of p

$$\begin{aligned}
&v_0(\xi,\eta) = e^{-\xi},\\
&p^1 v_1(\xi,\eta) = p^1 v_0(\xi,\eta) - p^1\pounds^{-1}\left[\frac{1}{s^\beta}\pounds\left\{s^\beta\frac{\partial v_0}{\partial\eta}(\xi,\eta) + \frac{\partial v_0}{\partial\xi} - 2\frac{\partial\partial^2 v_0}{\partial\eta\partial\xi^2}\right\}\right],\\
&v_1(\xi,\eta) = e^{-\xi} + e^{-\xi}\frac{\eta^\beta}{\Gamma(\beta+1)},\\
&p^2 v_2(\xi,\eta) = p^2 v_1(\xi,\eta) - p^2\pounds^{-1}\left[\frac{1}{s^\beta}\pounds\left\{s^\beta\frac{\partial v_1}{\partial\eta}(\xi,\eta) + \frac{\partial v_1}{\partial\xi} - 2\frac{\partial\partial^2 v_1}{\partial\eta\partial\xi^2}\right\}\right],\\
&v_2(\xi,\eta) = e^{-\xi} + e^{-\xi}\frac{\eta^\beta}{\Gamma(\beta+1)} + e^{-\xi}\frac{\eta^{2\beta}}{\Gamma(2\beta+1)},\\
&p^3 v_3(\xi,\eta) = p^3 v_2(\xi,\eta) - p^3\pounds^{-1}\left[\frac{-1}{s^\beta}\pounds\left\{s^\beta\frac{\partial v_2}{\partial\eta}(\xi,\eta) + \frac{\partial v_2}{\partial\xi} - 2\frac{\partial\partial^2 v_2}{\partial\eta\partial\xi^2}\right\}\right],\\
&v_3(\xi,\eta) = e^{-\xi} + e^{-\xi}\frac{\eta^\beta}{\Gamma(\beta+1)} + e^{-\xi}\frac{\eta^{2\beta}}{\Gamma(2\beta+1)} + e^{-\xi}\frac{\eta^{3\beta}}{\Gamma(3\beta+1)}.\\
&\quad\vdots
\end{aligned}$$

The analytical expression is therefore obtained in the following way

$$v(\xi,\eta) = e^{-\xi} + e^{-\xi}\frac{\eta^\beta}{\Gamma(\beta+1)} + e^{-\xi}\frac{\eta^{2\beta}}{\Gamma(2\beta+1)} + e^{-\xi}\frac{\eta^{3\beta}}{\Gamma(3\beta+1)} + \cdots. \tag{30}$$

If $\beta = 1$ the series form is

$$v(\xi,\eta) = e^{-\xi}\left(1 + \eta + \frac{\eta^2}{2!} + \frac{\eta^3}{3!} + \cdots\right) \tag{31}$$

The exact solution at $\beta = 1$

$$v(\xi,\eta) = e^{\eta-\xi}. \tag{32}$$

In Table 2, the analytical solutions of VHPTM and HPSTM are compared in terms of absolute error. The accuracy has been measured for both the methods. By comparison it has shown that the proposed method VHPTM has a higher degree of accuracy than HPSTM.

In Figure 5, the graphs of exact and approximate solutions of example 4.3 are plotted. The graphical representation has confirmed that exact and VHPTM solutions are coincident. The exact and approximate solutions are closed to each other and verify the validity of the proposed method. The solution of example 4.3 at different fractional-orders $\beta = 1, 0.8, 0.6$ and 0.4 are shown graphically in Figure 6. The obtained solutions support the convergence phenomena of the solution of fractional-order problems to the solution of integer-order problem for the example 4.3.

Table 2. Comparison of VHPTM and HPSTM [11] of example 4.3 at $\eta = 0.0000001$.

	VHPTM	Exact	Absolute Error	Absolute Error
ξ	$\beta = 1$		$HPSTM(\beta = 1)$	$VHPTM(\beta = 1)$
0.1	0.9048374271	0.9048374180	$8.396197700 \times 10^{-09}$	$9.1000000 \times 10^{-10}$
0.2	0.8187307613	0.8187307531	$2.343247660 \times 10^{-09}$	$8.2000000 \times 10^{-09}$
0.3	0.7408182281	0.7408182207	$1.098518064 \times 10^{-08}$	$7.4000000 \times 10^{-09}$
0.4	0.6703200527	0.6703200460	$2.435301980 \times 10^{-08}$	$6.7000000 \times 10^{-09}$
0.5	0.6065306658	0.6065306597	$3.925301546 \times 10^{-08}$	$6.1000000 \times 10^{-09}$
0.6	0.5488116416	0.5488116361	$5.222691016 \times 10^{-08}$	$5.5000000 \times 10^{-09}$
0.7	0.4965853088	0.4965853038	$6.061930396 \times 10^{-08}$	$5.0000000 \times 10^{-09}$
0.8	0.4493289686	0.4493289641	$6.351737693 \times 10^{-08}$	$4.5000000 \times 10^{-09}$
0.9	0.4065696638	0.4065696597	$6.159048138 \times 10^{-08}$	$4.1000000 \times 10^{-09}$
1	0.3678794449	0.3678794412	$5.612475460 \times 10^{-08}$	$3.7000000 \times 10^{-09}$

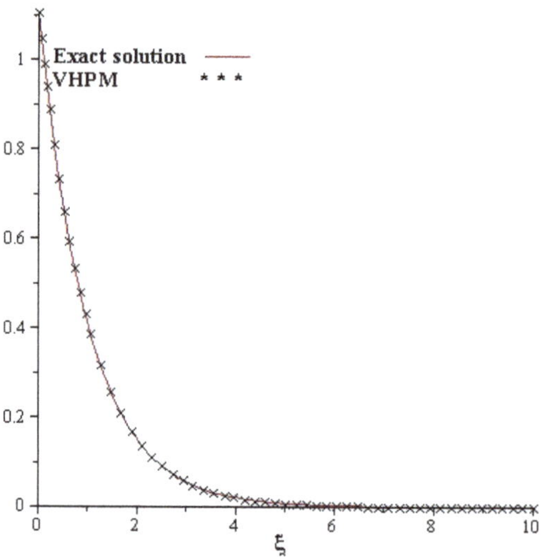

Figure 5. Variational homotopy perturbation transform method solution of example 4.3 for $\beta = 1$.

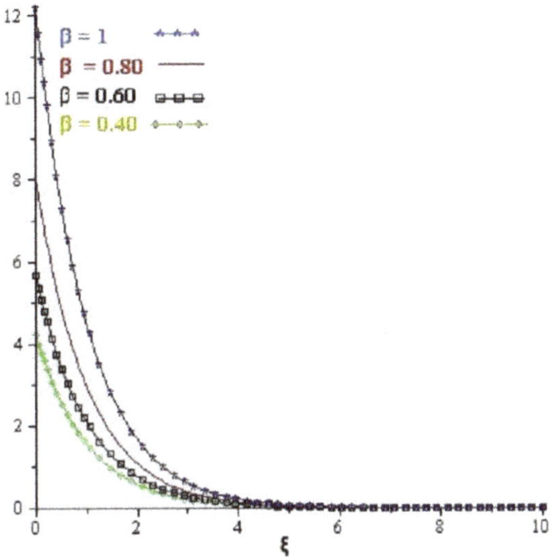

Figure 6. Variational homotopy perturbation transform method solution of example 4.3 at different fractional order $\beta = 1, 0.8, 0.6, 0.4$.

4.4. Example

We consider time fractional-order linear RLW equation

$$\frac{\partial^\beta v(\xi.\eta)}{\partial \eta^\beta} + \frac{\partial^4 v(\xi.\eta)}{\partial \xi^4} = 0, \qquad 0 < \xi \leq 1, \qquad 0 < \beta \leq 1, \quad \eta > 0, \tag{33}$$

with initial condition

$$v(\xi, 0) = \sin \xi, \tag{34}$$

By using Equation (12), the fractional PDE given in Equation (33) can be written as

$$\sum_{j=0}^{\infty} p^j v_{j+1}(\xi, \eta) = \sum_{j=0}^{\infty} p^j v_j(\xi, \eta) + \sum_{j=0}^{\infty} p^j £^{-1} \left[\lambda(s) £ \left\{ s^\beta \frac{\partial^\beta v_j(\xi.\eta)}{\partial \eta^\beta} (\xi, \eta) + \frac{\partial^4 v_j(\xi.\eta)}{\partial \xi^4} \right\} \right], \tag{35}$$

where $\lambda(s)$ is the Lagrange multiplier

$$\lambda(s) = \frac{-1}{s^\beta},$$

Applying VHPTM using He's polynomials,

$$\begin{aligned} \sum_{j=0}^{\infty} p^j v_j(\xi, \eta) = \sum_{j=0}^{\infty} p^j v_j(\xi, \eta) - p £^{-1} \left[\frac{1}{s^\beta} £ \left\{ s^\beta \left(\frac{\partial v_0}{\partial \eta} + p \frac{\partial v_1}{\partial \eta} + p^2 \frac{\partial v_2}{\partial \eta} + \cdots \right) \right.\right. \\ \left.\left. + \frac{\partial^4}{\partial \xi^4} \left\{ v_0 + p v_1 + p^2 v_2 + \cdots \right\} \right\} \right] \end{aligned} \tag{36}$$

Comparing the coefficients of p

$$v_0(\xi,\eta) = \sin\xi,$$
$$p^1 v_1(\xi,\eta) = p^1 v_0(\xi,\eta) - p^1 \pounds^{-1}\left[\frac{1}{s^\beta}\pounds\left\{s^\beta \frac{\partial v_0(\xi.\eta)}{\partial \eta}(\xi,\eta) + \frac{\partial^4 v_0(\xi.\eta)}{\partial \xi^4}\right\}\right],$$
$$v_1(\xi,\eta) = \sin\xi - \sin\xi\frac{\eta^\beta}{\Gamma(\beta+1)},$$
$$p^2 v_2(\xi,\eta) = p^2 v_1(\xi,\eta) - p^2 \pounds^{-1}[\frac{1}{s^\beta}\pounds\{s^\beta \frac{\partial v_1(\xi.\eta)}{\partial \eta}(\xi,\eta) + \frac{\partial^4 v_1(\xi.\eta)}{\partial \xi^4}\}],$$
$$v_2(\xi,\eta) = \sin\xi - \sin\xi\frac{\eta^\beta}{\Gamma(\beta+1)} + \sin\xi\frac{\eta^{2\beta}}{\Gamma(2\beta+1)},$$
$$p^3 v_3(\xi,\eta) = p^3 v_2(\xi,\eta) - p^3 \pounds^{-1}\left[\frac{1}{s^\beta}\pounds\left\{s^\beta \frac{\partial v_2(\xi.\eta)}{\partial \eta}(\xi,\eta) + \frac{\partial^4 v_2(\xi.\eta)}{\partial \xi^4}\right\}\right],$$
$$v_3(\xi,\eta) = \sin\xi - \sin\xi\frac{\eta^\beta}{\Gamma(\beta+1)} + \sin\xi\frac{\eta^{2\beta}}{\Gamma(2\beta+1)} - \sin\xi\frac{\eta^{3\beta}}{\Gamma(3\beta+1)}.$$

.

.

.

The analytical expression is therefore obtained in the following way

$$v(\xi,\eta) = \sin\xi - \sin\xi\frac{\eta^\beta}{\Gamma(\beta+1)} + \sin\xi\frac{\eta^{2\beta}}{\Gamma(2\beta+1)} - \sin\xi\frac{\eta^{3\beta}}{\Gamma(3\beta+1)} + \cdots. \tag{37}$$

If $\beta = 1$ the series form is

$$v(\xi,\eta) = \sin\xi\left(1 - \eta + \frac{\eta^2}{2!} - \frac{\eta^3}{3!} + \cdots\right). \tag{38}$$

The exact solution at $\beta = 1$

$$v(\xi,\eta) = \sin\xi e^{-\eta}. \tag{39}$$

In Figure 7, the exact and VHPTM solution for example 4.4 are plotted. It can be seen from the figure that exact and VHPTM solutions are in closed contact with each other. In Figure 8, the VHPTM solutions for the example 4.4 at different fractional-orders are calculated. The convergence of fractional-order solutions towards integer-order solution has proved the applicability of the proposed method.

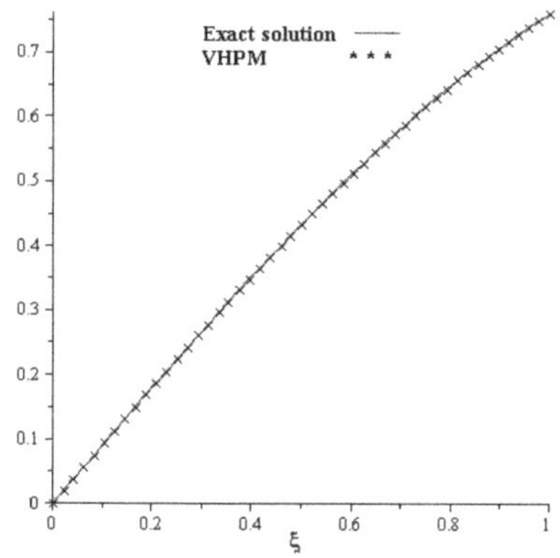

Figure 7. Variational homotopy perturbation transform method solution of example 4.4 for $\beta = 1$.

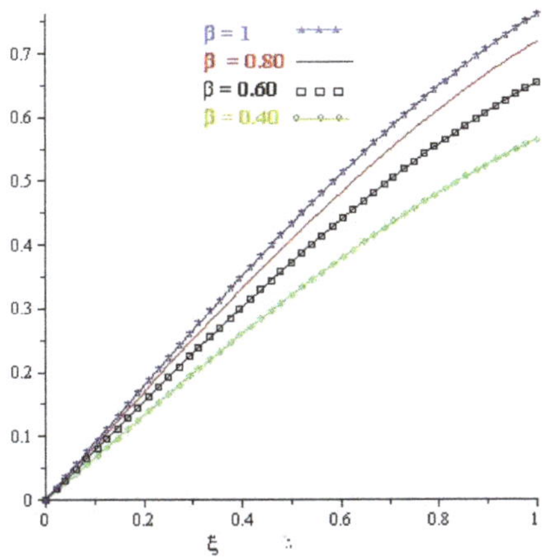

Figure 8. Variational homotopy perturbation transform method solution of example 4.4 at different fractional order $\beta = 1, 0.8, 0.6, 0.4$.

5. Results and Discussion

Several numerical examples are considered checking the applicability and reliability of the proposed method. The graphical representations of the solutions of examples 1 to 4, have provided the information about the accuracy and reliability of the proposed method. All the results of examples 1–4, confirmed strong agreement of VHPTM solutions with the exact solutions of the problems. The graphs represented the solutions for each problems at different fractional-order $\beta = 1, 0.8, 0.6$ and 0.4. Investigations show that solutions of fractional-order problem are convergent to the solution of integer-order problems. Moreover, the simple and straightforward implementation of the suggested method is also observed throughout the simulation. From the above properties of the present method, we expect that it can be modified for other fractional-order differential equations which arise in science and engineering.

6. Conclusions

In this article, the fractional view of acoustic wave equation is discussed by using a modified analytical technique. The solution graphs are plotted to provide clear pictures and analysis of the obtained results. The graphical representation has suggested the greatest rate of convergence as compared to other analytical methods. The fractional-order analysis of the acoustic wave equation is important to investigate the behaviour of the dynamics as compared to the classical one. Therefore, in the present application scenario, the proposed method has played a significant role to describe sophisticated solutions of fractional-order partial differential equations arising in different areas of sciences and engineering. Moreover, the present method uses the variational parameters which reduces the calculations' complexity. Also, the He's polynomials have been used to obtain the solutions in an accurate way as compared to Adomian polynomials. The rate of convergence of the suggested method is found to be higher than other existing methods. Hence, it is concluded that the present method can be extended to solve other fractional non-linear partial differential equations.

Author Contributions: Conceptualization, I.A. and H.K.; Methodology, R.S.; Software, H.K.; Validation, R.S. and I.A.; Formal Analysis, R.S.; Investigation, R.S. and D.B.; Resources, H.K. and P.K.; Writing—Original Draft Preparation, R.S.; Writing—Review and Editing, H.K., M.A. and P.K.; Visualization, M.A.; Supervision, D.B., P.K.; Project Administration, P.K.; Funding Acquisition, P.K. All authors have read and agreed to the published version of the manuscript.

Funding: Center of Excellence in Theoretical and Computational Science (TaCS-CoE) Faculty of Science, King Mongkuts University of Technology Thonburi (KMUTT).

Conflicts of Interest: The authors declare no conflict of interest.

References

1. Debnath, L. Recent applications of fractional calculus to science and engineering. *Int. J. Math. Math. Sci.* **2003**, *54*, 3413–3442. [CrossRef]
2. Miller, K.S.; Ross, B. *An Introduction to the Fractional Calculus and Fractional Differential Equations*; Wiley: New York, NY, USA, 1993.
3. Meerschaert, M.M.; Tadjeran, C. Finite difference approximations for two-sided space-fractional partial differential equations. *Appl. Numer. Math.* **2006**, *56*, 80–90. [CrossRef]
4. Parvizi, M.; Eslahchi, M.R.; Dehghan, M. Numerical solution of fractional advection-diffusion equation with a nonlinear source term. *Numer. Algorithms* **2015**, *68*, 601–629. [CrossRef]
5. Bu, W.; Liu, X.; Tang, Y.; Yang, J. Finite element multigrid method for multi-term time fractional advection diffusion equations. *Int. J. Model. Simul. Sci. Comput.* **2015**, *6*, 1540001. [CrossRef]
6. Donatelli, M.; Mazza, M.; Serra-Capizzano, S. Spectral analysis and structure preserving preconditioners for fractional diffusion equations. *J. Comput. Phys.* **2016**, *307*, 262–279. [CrossRef]
7. Donatelli, M.; Mazza, M.; Serra-Capizzano, S. Spectral analysis and multigrid methods for finite volume approximations of space-fractional diffusion equations. *SIAM J. Sci. Comput.* **2018**, *40*, A4007–A4039. [CrossRef]
8. Lin, X.L.; Ng, M.K.; Sun, H.W. A multigrid method for linear systems arising from time-dependent two-dimensional space-fractional diffusion equations. *J. Comput. Phys.* **2017**, *336*, 69–86. [CrossRef]
9. Shah, R.; Khan, H.; Kumam, P.; Arif, M. An Analytical Technique to Solve the System of Nonlinear Fractional Partial Differential Equations. *Mathematics* **2019**, *7*, 505. [CrossRef]
10. Khan, H.; Shah, R.; Baleanu, D.; Kumam, P.; Arif, M. Analytical Solution of Fractional-Order Hyperbolic Telegraph Equation, Using Natural Transform Decomposition Method. *Electronics* **2019**, *8*, 1015. [CrossRef]
11. Goswami, A.; Singh, J.; Kumar, D.; Gupta, S. An efficient analytical technique for fractional partial differential equations occurring in ion acoustic waves in plasma. *J. Ocean Eng. Sci.* **2019**, *4*, 85–99. [CrossRef]
12. Khan, Y.; Taghipour, R.; Falahian, M.; Nikkar, A. A new approach to modified regularized long wave equation. *Neural Comput. Appl.* **2013**, *23*, 1335–1341. [CrossRef]
13. Peregrine, D.H. Calculations of the development of an undular bore. *J. Fluid Mech.* **1966**, *25*, 321–330. [CrossRef]
14. Benjamin, T.B.; Bona, J.L.; Mahony, J.J. Model equations for long waves in nonlinear dispersive systems. *Philos. Trans. R. Soc. Lond. Ser. A Math. Phys. Sci.* **1972**, *272*, 47–78. [CrossRef]
15. Khalifa, A.K.; Raslan, K.R.; Alzubaidi, H.M. Numerical study using ADM for the modified regularized long wave equation. *Appl. Math. Model.* **2008**, *32*, 2962–2972. [CrossRef]
16. Achouri, T.; Omrani, K. Numerical solutions for the damped generalized regularized long-wave equation with a variable coefficient by Adomian decomposition method. *Commun. Nonlinear Sci. Numer Simul.* **2009**, *14*, 2025–2033. [CrossRef]
17. İdris, D. Least-squares quadratic B-spline finite element method for the regularised long wave equation. *Comput. Methods Appl. Mech. Eng.* **2000**, *182*, 205–215.
18. Bota, C.; Căruntu, B. Approximate analytical solutions of the regularized long wave equation using the optimal homotopy perturbation method. *Sci. World J.* **2014**, *2014*, 721865. [CrossRef]
19. Ganji, D.D.; Tari, H.; Jooybari, M.B. Variational iteration method and homotopy perturbation method for nonlinear evolution equations. *Comput. Math. Appl.* **2007**, *54*, 1018–1027. [CrossRef]
20. Inc, M.; Uğurlu, Y. Numerical simulation of the regularized long wave equation by He's homotopy perturbation method. *Phys. Lett. A* **2007**, *369*, 173–179. [CrossRef]
21. Noor, M.A.; Mohyud-Din, S.T. Modified variational iteration method for heat and wave-like equations. *Acta Appl. Math.* **2008**, *104*, 257–269. [CrossRef]
22. Noor, M.A.; Mohyud-Din, S.T. Variational homotopy perturbation method for solving higher dimensional initial boundary value problems. *Math. Probl. Eng.* **2008**, *2008*, 696734. [CrossRef]
23. Mohammad Mehdi Hosseini, S.; Tauseef Mohyud-Din, S.; Ghaneai, H. Variational iteration method for Hirota-Satsuma coupled KdV equation using auxiliary parameter. *Int. J. Numer. Methods Heat Fluid Flow* **2012**, *22*, 277–286. [CrossRef]

24. Mohyud-Din, S.T.; Yildirim, A.; Hosseini, M.M. Variational iteration method for initial and boundary value problems using He's polynomials. *Int. J. Differ. Equ.* **2010**, *2010*, 426213. [CrossRef]
25. Liu, Y. Variational homotopy perturbation method for solving fractional initial boundary value problems. In *Abstract and Applied Analysis*; Hindawi: London, UK, 2012; Volume 2012.
26. Shah, R.; Khan, H.; Baleanu, D. Fractional Whitham–Broer–Kaup Equations within Modified Analytical Approaches. *Axioms* **2019**, *4*, 125. [CrossRef]
27. Shah, R.; Khan, H.; Farooq, U.; Baleanu, D.; Kumam, P.; Arif, M. A New Analytical Technique to Solve System of Fractional-Order Partial Differential Equations. *IEEE Access* **2019**, *7*, 150037–150050. [CrossRef]
28. Shah, R.; Khan, H.; Arif, M.; Kumam, P. Application of Laplace-Adomian Decomposition Method for the Analytical Solution of Third-Order Dispersive Fractional Partial Differential Equations. *Entropy* **2019**, *21*, 335. [CrossRef]
29. Mahmood, S.; Shah, R.; Arif, M. Laplace Adomian Decomposition Method for Multi Dimensional Time Fractional Model of Navier-Stokes Equation. *Symmetry* **2019**, *11*, 149. [CrossRef]
30. Khan, H.; Shah, R.; Baleanu, D.; Arif, M. An Efficient Analytical Technique, for The Solution of Fractional-Order Telegraph Equations. *Mathematics* **2019**, *7*, 426. [CrossRef]
31. Shah, K.; Khalil, H.; Khan, R.A. Analytical solutions of fractional order diffusion equations by natural transform method. *Iran. J. Sci. Technol. Trans. A Sci.* **2018**, *42*, 1479–1490. [CrossRef]

MDPI
St. Alban-Anlage 66
4052 Basel
Switzerland
Tel. +41 61 683 77 34
Fax +41 61 302 89 18
www.mdpi.com

Applied Sciences Editorial Office
E-mail: applsci@mdpi.com
www.mdpi.com/journal/applsci

www.ingramcontent.com/pod-product-compliance
Lightning Source LLC
Chambersburg PA
CBHW062036220926
51850CB00010B/2567
9783039288250